Nuffield Maths 5
Teachers' Handbook

Nuffield Maths 5-11

a revised and extended version of
the Nuffield Mathematics Teaching Project

Nuffield Maths 5
Teachers' Handbook

Published for The Nuffield-Chelsea Curriculum Trust
by Longman Group Ltd

General editor:
Eric A. Albany,
Formerly Senior lecturer in mathematics education,
The Polytechnic, Wolverhampton

Associate editor:
Raymond J. Bull,
Formerly Senior lecturer in mathematics education,
The Polytechnic, Wolverhampton

Author of this volume:
John Hargreaves,
Warden of Morpeth Teachers' Centre, Northumberland

Contributory Authors:
Eric A. Albany,
Raymond J. Bull,
Ron Wyvill,
Formerly Headmaster,
Huish County Primary School, Yeovil

Illustrator:
Chris Williamson

We are grateful to:
Peter Chivers, Headmaster and Nigel Catley,
Deputy Headmaster and Class 1 of
Duxford C. E. Community School,
Duxford, Cambridge
for the cover photograph.

LONGMAN GROUP LIMITED
Longman House, Burnt Mill, Harlow, Essex, U.K.

First published 1982
ISBN 0 582 19179 3

Photoset in 10/12 Monophoto Plantin 110 and 194
by Keyspools Ltd. Golborne, Lancashire

Printed in Great Britain by
Hazell Watson & Viney Limited, Aylesbury

Contents

Foreword

As organizer of 'Nuffield Mark I', I am delighted to have the opportunity of welcoming the present publication, which in effect is Nuffield Mark II. The original project started in 1964 with the aim of 'producing a contemporary course', an urgent need at the time when the 11-plus examination in arithmetic was on its way out and there was a realization that neither its contents nor the methods of teaching for it were producing happy or numerate children (the overwhelming majority of people in that era grew up to hate and fear the subject).

The decision was taken at that time to produce only guidance for the teachers of primary children and not materials for the children themselves. Arguments will continue to rage as to whether this was a wise decision. I can defend it vigorously on many counts *at that time*, but I am also glad to be on record as saying that about ten years later there would be the acceptance and the need for the production of pupils' materials as well.

And so, of course, it has turned out. Very many teachers have asked for more guidance and more materials to be put into the hands of their pupils, and this is just what Mark II has set out to achieve. It is very fortunate that this enterprise has been directed by Eric Albany. He is a staunch Nuffield man who contributed a lot to Mark I. His ability, shrewdness and sense of both humour and proportion have ensured that Mark II should complete the task of 'producing a contemporary course' which children can enjoy. Of equal importance, they will be helped to think for themselves and to acquire relevant skills to the very best of their ability. Eric Albany and his team have produced materials which will set a standard of excellence for many years to come.

The work of the Nuffield Foundation in mathematics and science education has now been taken over by the Nuffield-Chelsea Curriculum Trust. Among the many institutions and people to whom the Trust owes thanks for their help, I must especially acknowledge the part played by the Polytechnic, Wolverhampton in allowing the full-time secondment of Eric Albany to the project and also the assistance given by Wolverhampton and Walsall Education Committees in providing accommodation and facilities for the project staff. We are extremely grateful to all those teachers and schools who have taken part in the trials of the new materials. I would also like to express our thanks to William Anderson, Publications Manager of the project and his colleagues, to the project secretary, Val Whitticase, and to our publishers, Longman Group Ltd, who have devoted so much effort and such skill to the editing, design and production of the materials.

Geoffrey Matthews

Chairman of the Nuffield-Chelsea Curriculum Trust
Primary Mathematics Consultative Committee
Professor Emeritus, Chelsea College,
University of London

Introduction

Nuffield Maths 5–11 is based on the original *Nuffield Mathematics Teaching Project* but is revised in the light of experience, and extended to include the full range of pupils' materials. *Nuffield Maths 5 Pupils' Book* is a *non-expendable* text intended for children approximately 9 to 10 years old.

Aims and objectives

The general aim of the *Nuffield Maths 5–11* Project is to promote understanding of the concepts and proficiency in the basic skills of mathematics in children of the 5–11 age range.

The objectives of the Teachers' Handbook are:
a) To give teachers clear guidance on the content, method and timing appropriate at each stage of the course;
b) To give practical, 'down to earth' suggestions for teaching Number, Measurement and Shape, using activities suitable for children with a wide range of abilities and backgrounds;
c) To give guidance in the use of both home-made and commercially available apparatus;
d) To encourage the development of a healthy, inquisitive attitude towards mathematical patterns and structures;
e) To suggest ways of dealing with children's difficulties.

Using the materials

The materials of the *Nuffield Maths 5–11* Project can be used in a variety of classroom organisations including individual work, group or class teaching. This should prove particularly useful to the teacher who tends to vary the type of organisation to suit particular topics. Whichever system is used, it is important for teachers to remember the following points:
a) Children learn at different rates and so will not reach the same stage simultaneously;
b) Young children learn by doing and by discussion;
c) As well as finding out and 'discovering' things about mathematics, children need to be *told* things about mathematics, particularly if new vocabulary is involved.

The obvious line of development for a primary child learning mathematics would seem to be:

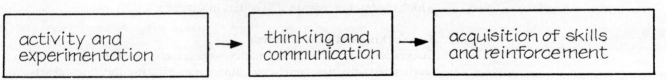

activity and experimentation → thinking and communication → acquisition of skills and reinforcement

Activity and experimentation may vary from a child 'finding out by fiddling' to a structured or teacher-led activity.

Thinking and communication involves discussion, sometimes between children, sometimes between teacher and children. Discussion leads to recording in various forms:
a) copying and completing statements – in this case, □ is intended to be replaced by a numeral so that □ + 6 = 15, for example, would be copied and completed as 9 + 6 = 15;
b) drawing or completing shapes or diagrams;
c) recording estimates and measurements;
d) using tabulations such as addition squares or investigating patterns on 100 squares.

Often children spend more time ruling up a table or chart than actually performing the calculations needed to complete it. To overcome this, *Nuffield Maths 5 Spiritmasters* are available from which copies can be run off ready for children's use.

Acquisition of skills and reinforcement   Apart from the obvious benefits of having useful skills and facts at one's fingertips, there is the question of building up confidence and enjoyment – 'I can do these. Can I have some more?'

The important thing is that these three elements form a *sequence*. The exercises in the Pupils' Book are seen as part of the last element of the sequence. Additional exercises, to be used as extra practice or for assessment, are available as part of the *Nuffield Maths 5 Spiritmasters* pack.

Chapter format

Each of the 21 chapters in the Teachers' Handbook is set out as follows:

1   For the teacher:
A brief outline of what is being attempted in the chapter, where it is leading, and what may need to be revised before starting.

2   Summary of the stages:
Setting out the stages contained within a chapter.

3   Vocabulary
A list of words and phrases which the children will need to be able to use and understand if they are to appreciate and explore the ideas in each chapter. The teacher may wish to include some of these words and phrases in work on language.

4   Equipment and apparatus
The sort of materials such as boxes, containers, pots, sticks, pictures, sorting toys, buttons, counters, cubes, beads, string, sand, plasticine, etc. which the teacher may need to collect in advance.

5   Working with the children
Suggestions for introducing and developing each stage through discussion, teacher-led activities, games, etc; hints for making number lines, charts,

models, displays and simple apparatus; how to check-up, where necessary, that a child understands a particular stage.

6   Pages from the Pupil's Book and Spiritmasters

These provide an invaluable link between teaching notes and pupils' material. *Answers* to specific and computational questions in the Pupils' Book are given on pages 122 to 130. The Spiritmasters provide extra practice material.

7   References and resources

A list of books and commercially produced materials which are appropriate for the chapter. No commercially produced equipment is deemed essential but is suggested as a possible alternative to homemade or environmental materials. Occasionally the Teachers' Guides published by the original *Nuffield Foundation Mathematics Teaching Project* may be listed in this section. These guides are now out of print but may be found in libraries and schools and they still make a valuable contribution to mathematics education.

Addition

For the teacher

Once the thousand has been introduced properly, using apparatus and diagrammatic representations, then the addition of four-digit numbers involves the same basic processes as those described earlier. This chapter follows closely the format and approach used previously for the addition of tens and units and of hundreds, tens and units. Apart from the 'thousands' column, nothing new is being brought in; we are merely making a natural extension of the number system.

The types of apparatus suggested include multibase arithmetic blocks and tokens which are seen as an extension of the 'home-made' apparatus used previously. These token pieces cut from different coloured cards to represent thousands, hundreds, tens and units need not be of sizes which portray the exact numerical relationships. The important thing is that the tokens are placed in appropriate columns.

Summary of stages

1 Introducing the thousand
2 Adding thousands, hundreds, tens and units:
   a) Using apparatus
   b) Extended notation – horizontal and/or vertical layout
   c) Conventional layout
3 Word problems

Vocabulary

Row, column, sum, total, exchange, abacus, blocks, squares, longs, units, tokens, horizontal, vertical, odometer, partial sum.

Equipment and apparatus

Base 10 blocks, squares, longs and units, token pieces cut from coloured card, counting boards (home-made), coloured card and paper fasteners, dice.

Working with the children

1 Introducing the thousand
The dramatic, 'shunting' effect of adding 1 to 999 should be shown to the children using multibase apparatus or token pieces on a 'counting board' with clearly labelled columns. Discussion of the three-fold exchange involved should bring out the following points:

A thousand is ten hundreds $1\,000 = 10(100)$
or nine hundreds and ten tens $1\,000 = 9(100) + 10(10)$
or nine hundreds and nine tens
   and ten units $1\,000 = 9(100) + 9(10) + 10(1)$

Children should also be reminded of what happens to the odometer when 'the noughts come up'.

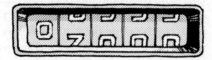

The abacus is used as a recording device in order to put particular emphasis on the *placing* of a digit in the correct *column*; it is **not** intended as a computing tool. The abacus also serves as a visual link between the digital and word forms of a number. (*Nuffield Maths 5 Spiritmasters*, Grid 30.)

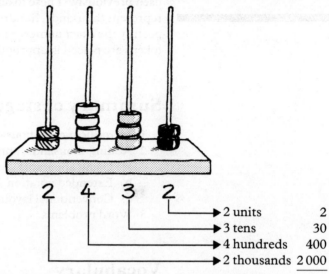

→ 2 units	2
→ 3 tens	30
→ 4 hundreds	400
→ 2 thousands	2 000
'two thousand four hundred and thirty-two'	2 432

It may be necessary to remind the children again of the use of a hyphen for the numerals twenty-one to ninety-nine.

Adding one, and later ten, to a given number provides extra opportunities for thinking about 'how a number is made up' – particularly if regrouping is necessary as in 1 094 + 10, for example.

The thousand marker

A comma should **not** be used as a thousand marker since in most European countries it is used as a decimal marker. On the continent 3,450 litres, for example, does not signify 'over three thousand litres' but means three litres and 450 millilitres' – the difference between drowning and quenching a thirst!

The Metrication Board in *How to Write Metric* says:

Four digits should be written without a space unless they form part of a tabulation,	3000

and then goes on:

Five or more digits may be grouped in blocks of three divided by single spaces,	3 000 000

and:

In tabulating columns of numbers,	2 000
the digits may be grouped in threes	19 000
from the decimal point.	425 321

The trouble with these recommendations is that when the first number, say 2 000, is written we may not always know whether subsequent numbers are going to contain more than 4 digits or not. This could lead to confusion:

$$
\begin{array}{r}
2\,000 \\
14\,262 \\
+\ \ 3\,188 \\
\hline
\end{array}
$$

Teachers may prefer to be more consistent and to encourage the use of a *small* space between groups of three digits in *all* calculations:

$$
\begin{array}{r}
3\,452 \\
+\,1\,876 \\
\hline
\end{array}
$$

2 Adding thousands, hundreds, tens and units

a Using apparatus

The counting board should now have four columns. It may help some children if, initially, the columns are labelled both 'thousands, hundreds, tens, units' and 'blocks, squares, longs, units' when base 10 pieces are used. An extra heading strip can be used for this.

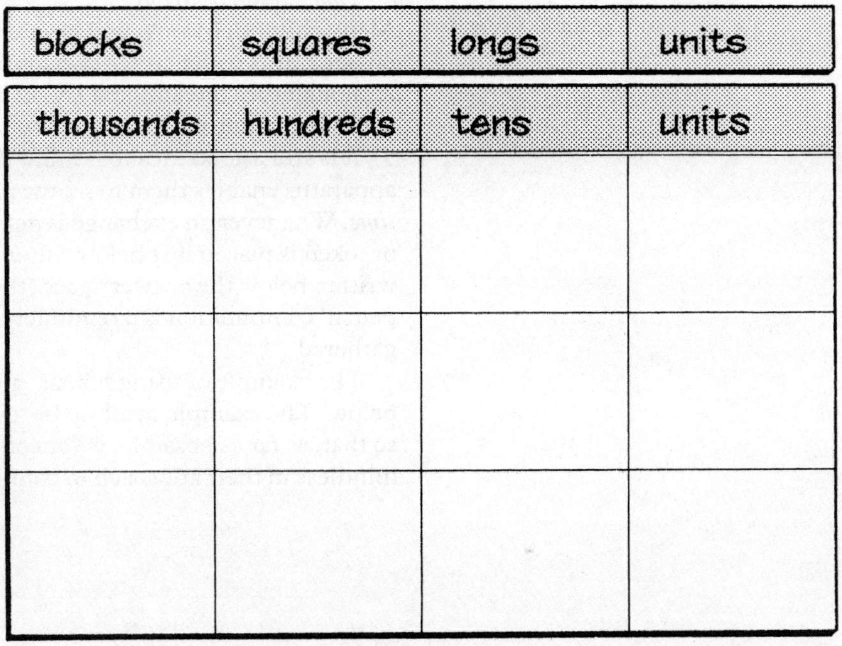

blocks	squares	longs	units
thousands	hundreds	tens	units

It is important that apparatus (either base 10 blocks, squares, longs and units or token cards) is used for some of the early exercises. The correct grouping and placing of the pieces reinforces previous work and strengthens the child's understanding of place value. It cannot be emphasised too strongly that acquiring the important concept of place value is a long, gradual process.

b Extended notation – horizontal and/or vertical layout

As before, extended notation is used as a link between what happens when apparatus is used and what is recorded on paper. Some teachers may consider that it is not necessary to go through *both* the horizontal *and* the vertical layouts. The fact that they are really only slightly different ways of setting-out is illustrated by the side-by-side comparison shown in *Nuffield Maths 5 Pupils' Book*.

Horizontal layout

$$2\,378\longrightarrow 2\,000 + 300 + 70 + 8$$
$$+\,4\,149\longrightarrow 4\,000 + 100 + 40 + 9$$
$$6\,000 + 400 + 110 + 17$$
$$6\,527 \qquad 6\,000 + 500 + 20 + 7$$

Vertical layout

2 378	
+ 4 149	
17	(8 + 9)
110	(70 + 40)
400	(300 + 100)
6 000	(2 000 + 4 000)
6 527	

The vertical layout has the advantage of being closer to the conventional one and avoids the final horizontal addition of the 'partial sums' which some children find difficult. On the other hand, the horizontal arrangement has the advantage of showing exactly how the 'partial sums' were obtained. However, this can be indicated initially in the vertical version by writing in brackets beside each partial sum the numbers which led to it.

c   Conventional layout

Many children by now will be adding four-digit numbers without relying on apparatus using either horizontal or vertical extended notation. Although the conventional layout has been used previously for two or three-digit addition (*Nuffield Maths 4 Teachers' Handbook*, Chapters 2 and 5), it is still a good idea to remind the children of what is happening. Using apparatus enables them to *see* the process of dealing with *one column at a time*. Whenever an exchange is necessary the new, correctly-labelled piece or token is placed just below the answer in the appropriate column. The 'I' written below the answer space (the 'carrying figure') in the 'paper-and-pencil' computation is a reminder of the piece or token waiting to be gathered.

The example of using tokens, given in the *Pupils' Book*, is illustrated below. The example needs to be 'talked through' with a group of children so that, as far as possible, we encourage children to be confident but not mindless in their approach to computation.

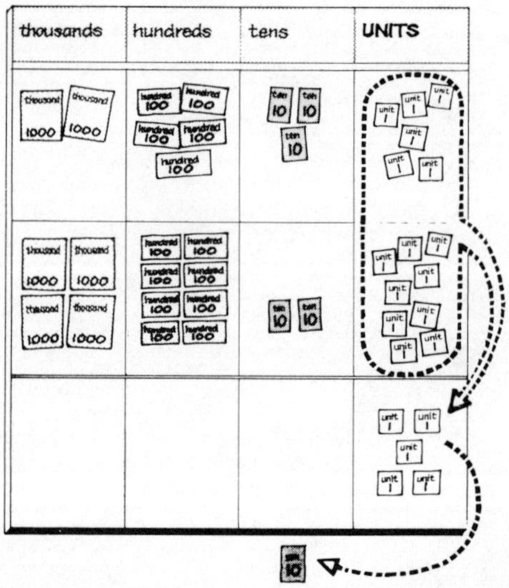

Units are scooped together. Ten units are exchanged for a ten-token to be put just below the answer box in the tens column. Five units left in answer.

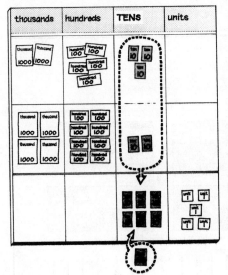

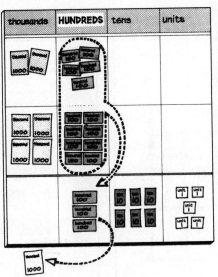

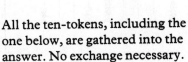

All the ten-tokens, including the one below, are gathered into the answer. No exchange necessary.

Hundred-tokens are scooped together. Ten hundreds are exchanged for one thousand-token to be put below the answer box in the *thousands* column. Three hundreds left in answer.

All the thousand-tokens, including the one below, gathered into answer.

$$
\begin{array}{r}
2\,536 \\
+\,4\,829 \\
\hline
7\,365 \\
\hline
1\quad 1
\end{array}
$$

3   Word problems

Children need reminding constantly about the importance of putting figures into the correct columns, especially when translating from the word-forms of numerals. Teachers will need to have a firm policy on the question of the thousand marker mentioned earlier. Problems involving dates now come within the scope of four-digit addition. 'My Grandad was born in 1927. In what year will he start to draw his old age pension?' Children should be encouraged to make up their own problems using situations from the environment.

An adaptation of the 'Dial-a-Sum' on page 32 of *Nuffield Maths 4 Teachers' Handbook*, to include four-digit numbers, will provide additional practice.

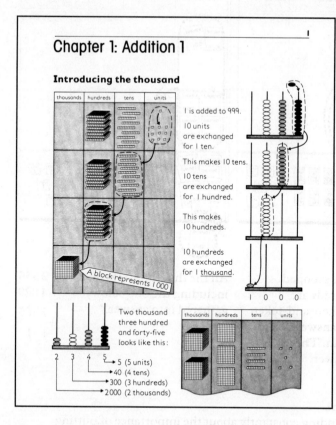

Chapter 1: Addition 1

Introducing the thousand

1 is added to 999.

10 units are exchanged for 1 ten.

This makes 10 tens.

10 tens are exchanged for 1 hundred.

This makes 10 hundreds.

10 hundreds are exchanged for 1 thousand.

A block represents 1000

Two thousand three hundred and forty-five looks like this:

5 (5 units)
40 (4 tens)
300 (3 hundreds)
2000 (2 thousands)

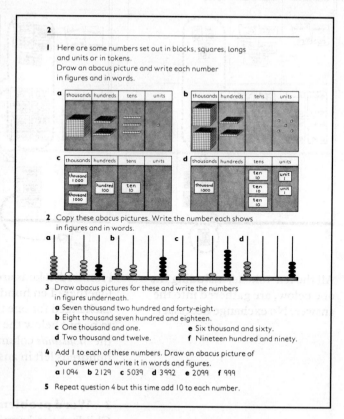

1 Here are some numbers set out in blocks, squares, longs and units or in tokens.
Draw an abacus picture and write each number in figures and in words.

a, b, c, d

2 Copy these abacus pictures. Write the number each shows in figures and in words.

a, b, c, d

3 Draw abacus pictures for these and write the numbers in figures underneath.
a Seven thousand two hundred and forty-eight.
b Eight thousand seven hundred and eighteen.
c One thousand and one.   e Six thousand and sixty.
d Two thousand and twelve.   f Nineteen hundred and ninety.

4 Add 1 to each of these numbers. Draw an abacus picture of your answer and write it in words and figures.
a 1094 b 2129 c 5039 d 3992 e 2099 f 999

5 Repeat question 4 but this time add 10 to each number.

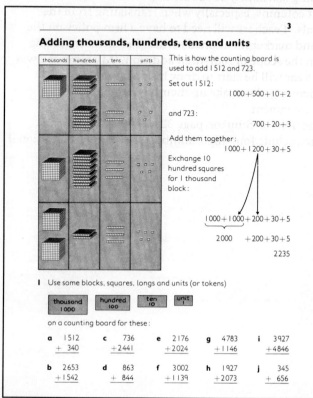

Adding thousands, hundreds, tens and units

This is how the counting board is used to add 1512 and 723.

Set out 1512:
$$1000+500+10+2$$

and 723:
$$700+20+3$$

Add them together:
$$1000+1200+30+5$$

Exchange 10 hundred squares for 1 thousand block:
$$1000+1000+200+30+5$$
$$2000+200+30+5$$
$$2235$$

1 Use some blocks, squares, longs and units (or tokens)

| thousand 1000 | hundred 100 | ten 10 | unit 1 |

on a counting board for these:

a 1512   c 736   e 2176   g 4783   i 3927
   + 340   +2441   +2024   +1146   +4846

b 2653   d 863   f 3002   h 1927   j 345
   +1542   + 844   +1139   +2073   + 656

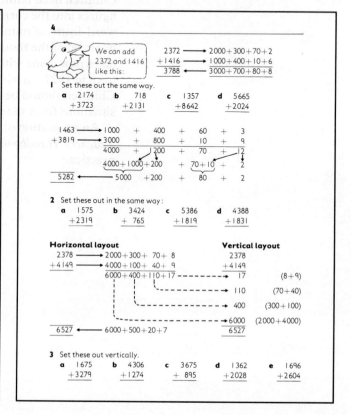

We can add 2372 and 1416 like this:

$$2372 \rightarrow 2000+300+70+2$$
$$+1416 \rightarrow 1000+400+10+6$$
$$3788 \leftarrow 3000+700+80+8$$

1 Set these out the same way.
a 2174   b 718   c 1357   d 5665
   +3723   +2131   +8642   +2024

$$1463 \rightarrow 1000 + 400 + 60 + 3$$
$$+3819 \rightarrow 3000 + 800 + 10 + 9$$
$$4000 + 1200 + 70 + 12$$
$$4000+1000+200 + 70+10 + 2$$
$$5282 \leftarrow 5000 + 200 + 80 + 2$$

2 Set these out in the same way:
a 1575   b 3424   c 5386   d 4388
   +2319   + 765   +1819   +1831

Horizontal layout

$$2378 \rightarrow 2000+300+ 70+ 8$$
$$+4149 \rightarrow 4000+100+ 40+ 9$$
$$6000+400+110+17$$
$$6527 \leftarrow 6000+500+20+7$$

Vertical layout

2378
+4149
17   (8+9)
110   (70+40)
400   (300+100)
6000   (2000+4000)
6527

3 Set these out vertically.
a 1675   b 4306   c 3675   d 1362   e 1696
   +3279   +1274   + 895   +2028   +2604

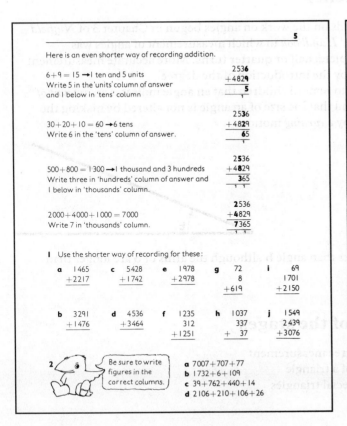

5

Here is an even shorter way of recording addition.

$6+9=15 \rightarrow 1$ ten and 5 units
Write 5 in the 'units' column of answer
and 1 below in 'tens' column.

```
 2536
+4829
    5
    1
```

$30+20+10=60 \rightarrow 6$ tens
Write 6 in the 'tens' column of answer.

```
 2536
+4829
   65
   1
```

$500+800=1300 \rightarrow 1$ thousand and 3 hundreds
Write three in 'hundreds' column of answer and
1 below in 'thousands' column.

```
 2536
+4829
  365
  1 1
```

$2000+4000+1000=7000$
Write 7 in 'thousands' column.

```
 2536
+4829
 7365
 1 1 1
```

1 Use the shorter way of recording for these:

a 1465	**c** 5428	**e** 1978	**g** 72	**i** 69	
+2217	+1742	+2978	8	1701	
			+619	+2150	

b 3291	**d** 4536	**f** 1235	**h** 1037	**j** 1549
+1476	+3464	312	337	2439
		+1251	+ 37	+3076

2 Be sure to write figures in the correct columns.

a $7007+707+77$
b $1732+6+109$
c $39+762+440+14$
d $2106+210+106+26$

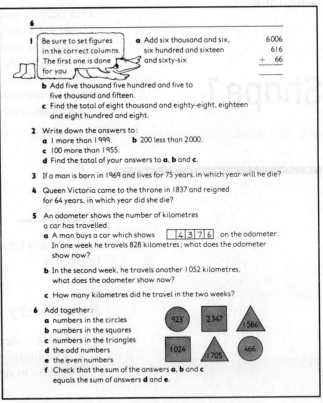

6

1 Be sure to set figures in the correct columns. The first one is done for you.

a Add six thousand and six, six hundred and sixteen and sixty-six

```
6006
 616
+ 66
```

b Add five thousand five hundred and five to five thousand and fifteen.

c Find the total of eight thousand and eighty-eight, eighteen and eight hundred and eight.

2 Write down the answers to:
a 1 more than 1999. **b** 200 less than 2000.
c 100 more than 1955.
d Find the total of your answers to **a**, **b** and **c**.

3 If a man is born in 1969 and lives for 75 years, in which year will he die?

4 Queen Victoria came to the throne in 1837 and reigned for 64 years, in which year did she die?

5 An odometer shows the number of kilometres a car has travelled.
a A man buys a car which shows ⟨4 3 7 6⟩ on the odometer. In one week he travels 828 kilometres; what does the odometer show now?
b In the second week, he travels another 1052 kilometres, what does the odometer show now?
c How many kilometres did he travel in the two weeks?

6 Add together:
a numbers in the circles
b numbers in the squares
c numbers in the triangles
d the odd numbers
e the even numbers
f Check that the sum of the answers **a**, **b** and **c** equals the sum of answers **d** and **e**.

Shapes: 923 (circle), 2347 (square), 1586 (triangle), 1024 (square), 1705 (triangle), 466 (circle)

**Practice material
for use with Nuffield Maths 5 Pupils' Book**

Practice 1
Chapter 1: Addition

Name _____

A

3216	5271	4279	6473	4693	1849	3279
+1485	+1853	+3684	+2874	+3478	+6764	+4962
4701	7124	7963	9347	8171	8613	8241

B

Shapes: 356, 1531, 2491, 1092, 1764, 2876, 848, 697, 2317, 1863, 3389, 4312

Find the total of the numbers in the:

rectangles	5095
hexagons	8567
triangles	5178
circles	4696

C
This table shows the numbers of passengers sailing to France on three cross-Channel ferries. Complete the totals first for each ship and then for each day.

	Princess of Wales	Shakespeare	Dover Castle	Totals for day
Monday	1574	1339	1619	4532
Tuesday	1735	1627	1922	5284
Wednesday	1827	1784	1888	5499
Thursday	1639	1588	1714	4941
Friday	1894	1769	2246	5909
Totals	8669	8107	9389	26165

Nuffield Maths 5 Spiritmasters   *Nuffield Maths 5 Pupils' Book, pages 1–6*   **14**

References and resources

Dienes, Z. P. *Building up Mathematics*, Hutchinson 1960

Williams, E. M. and Shuard, H. *Primary Mathematics Today* Third Edition (Chapter 13), Longman Group Ltd 1982

Arnold, E. J. *Tillich's Base 10 Blocks*

Classmate Triman, *Dial-a-Sum, Decimal Abacus*

E.S.A. *Base 10 Materials*

Nicholas Burdett *Base 10 Materials*

Taskmaster Ltd, *Decimetric Board*

Chapter 2

Shape 1

For the teacher

This chapter builds on the work on angles begun in Chapter 3 of *Nuffield Maths 4 Teachers' Handbook* in which measurement of angles was restricted to complete, half or quarter turns. More accurate measurement is made possible by the introduction of the degree.

It is advisable to remind children that an angle indicates an *amount of turn or rotation* and that the size of an angle is not altered by making the arms longer but by a *turning* motion.

 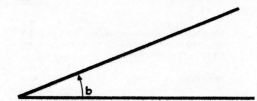

Angle **a** is larger than angle **b** although the arms of **b** are longer than those of **a**.

Summary of the stages

1 Angles – degree measurement
2 The angles of a triangle
3 Angles in special triangles

Vocabulary

Degree, clockwise, anti-clockwise, complete turn, right angle, acute, obtuse, right-angled, equilateral, isosceles, diagonal, north-east (NE), south-east (SE), south-west (SW), north-west (NW).

Equipment and apparatus

Paper circles, clock faces (rubber stamps), stiff paper or thin card, scissors.

Working with the children

1 Angles – degree measurement

Expressing a complete turn as 360°, half a complete turn as 180° and a quarter of a turn or right angle as 90° leads to an exercise similar to that in *Pupils' Book 4* but this time the amount of turn is recorded in degrees. Also, the inclusion of the extra four points on the compass, NE, SE, SW, and NW, brings in half right angles (45°) and multiples of 45°: 90°, 135°, 180°, 225°, 270°, 315°. (*Nuffield Maths 5 Spiritmasters*, Grid 1.)

On the clock face the 360° are divided into 12 equal sectors so a hand moving from one numeral to the next turns through 30°. The multiples of 30° involved in this exercise are: 30°, 60°, 90°, 120°, 150°, 180°, 210°, 240°, 270°, 300°, 330°. (*Nuffield Maths 5 Spiritmasters*, Grids 2 and 3.)

Right angles, acute ('sharp') angles and obtuse ('blunt') angles are revised but this time they are described in terms of degrees rather than fractions of a turn.

By this time children should be ready to estimate angles. This can be done easily if each child is given or makes two different coloured circles of paper cut along a radius and interleaved as shown in *Nuffield Maths 4 Teachers' Handbook*:

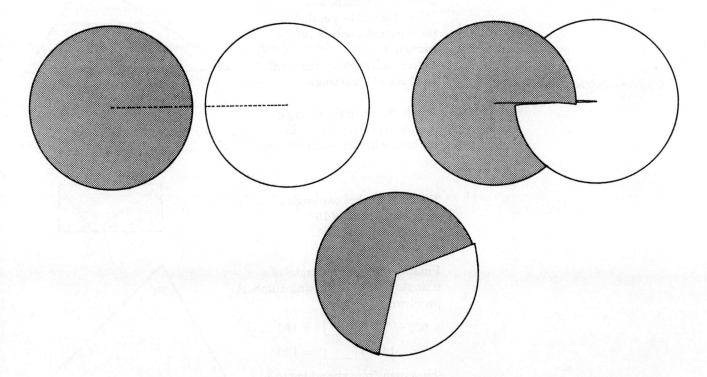

If the teacher then asks for an angle of, say 30°, to be made by rotating the disc and then held up, a quick visual check can be made.

Alternatively, a *rotogram* can be used (See References and Resources Section at the end of this chapter.)

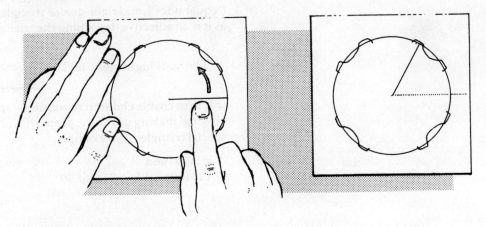

Being transparent, the **rotogram** has the advantage that it can be laid over another angle for comparison.

2 Angles of a triangle

As an alternative to the 'tearing and sticking' method shown in the *Pupils' Book*, a more permanent model can be made by folding:

Draw and cut out a triangle.
Find the midpoints of two sides
by measuring.
Join the two midpoints.
(This line will be parallel to
the third side, or base, of the
triangle.)
Letter each angle on the front
and back of the triangle.

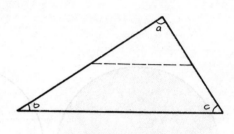

When the top of the triangle
is folded down along the line,
the apex will just touch the base.

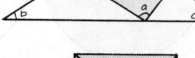

Fold in the other two angles
to show the total of 180°.

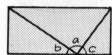

Finding the 'missing angle' of a
triangle is really a 'missing number'
problem.

$$62° + 73° + \boxed{}° = 180°$$
$$135° + \boxed{}° = 180°$$

'How many must be added to 135
to make 180?'

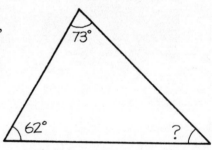

3 Angles in special triangles

Some children will need reminding of *isosceles* ('equal legs'), *equilateral* ('equal sides'), and *right-angled* triangles. Here right-angled is hyphenated as it is an adjective describing the triangle.

The marking of right angles, ⌐|, and equal sides

will then enable children to continue 'angle chasing' making use of the properties of special triangles. For example:

p is 75° and
r is 180° − (75° + 75°) or 30°

Pages from the Pupils' Book and Spiritmasters

7

Chapter 2: Shape 1

Angles-degree measurement

Using very simple instruments the Babylonians, before 4000 B.C., had the mistaken idea that it took the sun 360 days to complete its full journey round the earth. Because of this they divided a complete turn into 360 parts and called each part **a degree (written as 1°).**

One complete turn = 360°.
Half a complete turn = 180°.
A quarter of a complete turn or one right-angle = 90°.

1 The angle between N and E measured in a clockwise direction is a right angle. How many degrees is it?

2 The angle between N and NE measured in a clockwise direction is half a right angle. How many degrees is it?

3 The angle between N and SW measured in an anti-clockwise direction is 1½ right angles. How many degrees is it?

4 Copy and complete

I start facing	I turn through	Direction	I finish facing
North	180°	Clockwise	
South	90°	Anti-clockwise	
East	360°	Clockwise	
West	135°	Clockwise	
North East	225°		South
South West		Clockwise	South East
North West	270°		North East
South East		Clockwise	North
South East	315°		South
South East		Anti-clockwise	West

8

1 How many degrees does the minute hand of a clock turn through in one hour?

2 How long does it take the hour hand to turn through 360°?

3 How many degrees does the minute hand turn through in a quarter of an hour?

4 How many degrees does the minute hand turn through in 5 minutes?

5 In three hours the hour hand turns through how many degrees?

6 How many degrees does the hour hand turn through in one hour?

7 Copy and complete:

Hand of clock	From	To	Turns through (degrees)
hour hand	1	4	
minute hand	3	6	
minute hand	2	8	
hour hand	5	11	
minute hand	4	8	
hour hand	6	11	
minute hand	11	4	
hour hand	10	5	

Describing angles

A **right angle** = 90°

An **acute angle** is less than a right angle so an acute angle is less than 90°.

An **obtuse angle** is more than one right angle but less than two right angles so an obtuse angle is more than 90° but less than 180°.

8 List these angles and say whether they are acute or obtuse.

a 36°	**c** 79°	**e** 163°
b 141°	**d** 88°	**f** 17°

g 110°   **i** 34°   **k** 21°
h 96°   **j** 175°   **l** 169°

9

Angles of a triangle

1 Draw any triangle on a piece of paper. Mark the angles **A**, **B** and **C** as in the diagram.

Tear off each angle as shown:

Stick the three angles in your book like this:

Together they make 180° or two right angles.

> The angles of a triangle add up to 180°

2 For each triangle calculate the size of the unmarked angle:

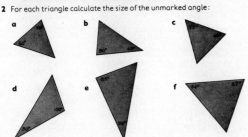

10

Angles in special triangles

A triangle with one angle of 90° is called a **right-angled triangle.** The right angle is often marked as in the diagram.

A triangle with its three sides equal is called an **equilateral triangle**. It also has three equal angles.

1 How many degrees is each angle?

A triangle with two equal sides is an **isosceles triangle**. It has two equal angles.

2 If one of the equal angles of an isosceles triangle is 55°, what are the sizes of the other two angles in the triangle?

Draw a square and the diagonal shown in the diagram.

3 How many degrees is the angle marked **a**?

4 How many is the angle marked **b**?

5 How many is the angle marked **c**?

6 Describe the triangle shaded in the diagram in two different ways.

7 For each triangle, calculate the size of the angles marked with letters.

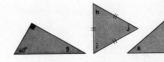

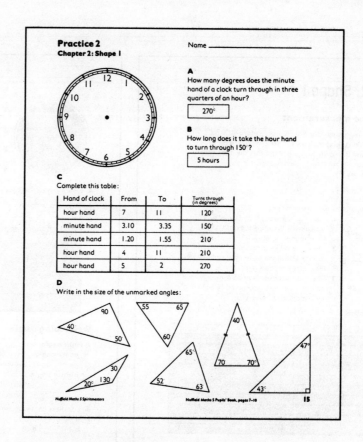

References and resources

Giles, G. *Rotagram I : Equal Angles* Booklets and Worksheets (Dime Project), Oliver & Boyd

Nuffield Mathematics Teaching Project, *Shape and Size* ▽, Nuffield Guide, Chambers/Murray, 1967 (See Introduction page xi.)

Williams, E. M. and Shuard, H. *Primary Mathematics Today* Third Edition (Chapters 5 and 20), Longman Group Ltd 1982

Dime project *Rotagram (Packets of 10)*.

Dime materials available from:
Geoffrey Giles, Dime, Dept of Education, University of Stirling, Stirling, Scotland

Subtraction

For the teacher

The extension of subtraction to deal with four-digit numbers does not require any new processes except the splitting of a thousand into ten hundreds. Although the children have already dealt with both extended notation and the shorter, traditional layout for three-digit subtraction, a brief return to the use of apparatus is made in order to remind them what the 'paper-and-pencil' computation represents. This is in line with the sequence suggested in *Nuffield Maths 4 Teachers' Handbook* (Chapter 12).

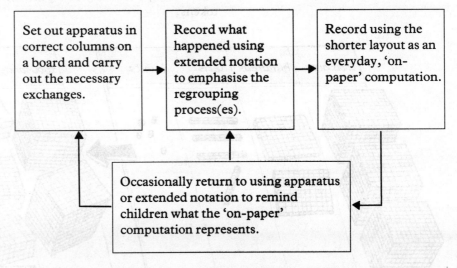

Set out apparatus in correct columns on a board and carry out the necessary exchanges. → Record what happened using extended notation to emphasise the regrouping process(es). → Record using the shorter layout as an everyday, 'on-paper' computation.

Occasionally return to using apparatus or extended notation to remind children what the 'on-paper' computation represents.

Alongside the development of decomposition as a written method for subtraction, it is as well to bear in mind that complementary addition is frequently used in practice, particularly when giving change or when 'counting on' from a smaller to a larger amount to 'find the difference'.

Summary of the stages

1 Subtraction of four-digit numbers (splitting thousands); recording by extended notation, shorter layout
2 Finding the difference by counting on
3 Word problems

Vocabulary

Subtract, take away, difference, minus, regroup(ing), exchange, not enough, represents, token.

Equipment and apparatus

Blocks, squares, ten-rods and units from Multibase arithmetic blocks (M.A.B) apparatus, coloured card for tokens (as an alternative), large sheets of card ruled for counting boards.

Working with the children

1 Subtraction of four-digit numbers (splitting thousands)

After reminding the children that a thousand is 'ten hundred', or
$1000 = 10(100)$, the four-column counting board is used with base 10
apparatus (either m.a.b. blocks, squares, longs and units or token cards) for
the first few examples. Besides giving a clear indication of what the abstract
'on paper' computation represents, grouping and placing pieces or tokens
in the correct columns continues to reinforce the gradually acquired place
value concept. Actually doing the exchanging and replacing of pieces or
tokens:

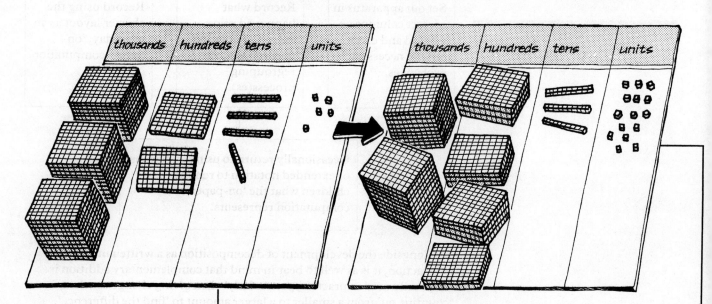

gives much more purpose and meaning to what might otherwise appear to
be mere juggling of figures:

$$(3000 + 200 + 40 + 5) \longrightarrow (2000 + 1200 + 30 + 15)$$

It is worth repeating that this extended notation represents the actual
disposition of the pieces and serves as a link between the concrete,
'counting board' version and the final, abstract 'on-paper' version:

$$(3000 + 200 + 40 + 5) \dashrightarrow \begin{array}{|c|c|c|c|} \hline 2 & 12 & 3 & 15 \\ \hline 3 & 2 & 4 & 5 \\ \hline \end{array} \leftarrow (2000 + 1200 + 30 + 15)$$

The extra problem of regrouping right across all four columns is
highlighted by asking children to record the exchange:

$$4000 = \boxed{3}(1000) + \boxed{9}(100) + \boxed{9}(10) + \boxed{10}(1)$$

Some children may need to be convinced of the equality of 1000 and (900 + 90 + 10), perhaps by building a 1000 block using nine 'layers' of 100 each and making up the top 'layer' with nine 10 rods and ten units.

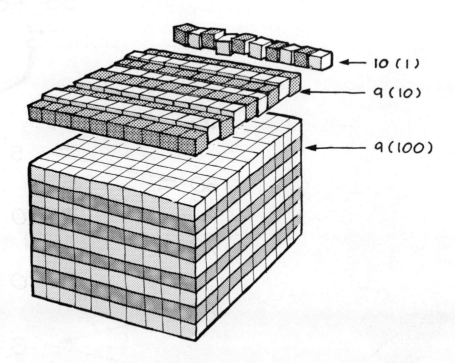

10 (1)

9 (10)

9 (100)

Alternatively, the inconvenience of going shopping with just one £10 note (worth 1000 pence) could be overcome by exchanging it for:

9 one-pound notes, 9 tenpence pieces and 10 pence.

This example again emphasises that it is a 'fair exchange', since both sums of money have the same value. It is just that the second arrangement is more convenient.

2 Finding the difference by counting on

Nuffield Maths 4 and the first stage of this chapter concentrated on developing decomposition as the 'paper-and-pencil' method for subtraction. Earlier, complementary addition or 'making up the difference' was used. To quote from *Nuffield Maths 3 Teachers' Handbook*:

This way of tackling subtraction is used everyday in shops to give 'change' – that is the difference between the price and the money offered – but it becomes the 'Cinderella' method in schools once larger numbers are involved. This may be because there is no concise form of recording it. This is a pity because 'making up' is very practical, is easily understood and helps to show the strong connection between subtraction and addition.

The children were introduced to two ways of recording, for example:

This could be called the 'candelabra' layout.

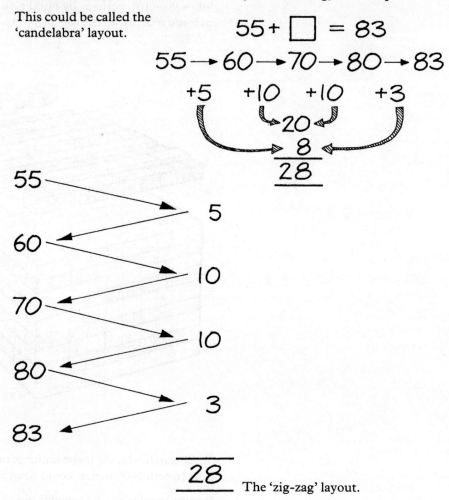

$$55 + \square = 83$$

$$55 \rightarrow 60 \rightarrow 70 \rightarrow 80 \rightarrow 83$$

+5   +10   +10   +3

20

8

$$\overline{28}$$

55

5

60

10

70

10

80

3

83

$$\overline{28}$$ The 'zig-zag' layout.

Of these, the zig-zag layout is probably easier when 4-digit numbers are involved.

It is possible that some children may be able to use the 'making up' method mentally. This means working out the size of each 'jump' (possibly by using a 'mental number line') while keeping a cumulative total of the jumps.

For example: $17 + \boxed{} = 45$

Imagine:

17   20   30   40   45

3   +10   +10   +5

Think: and 10   and 10   and 5

Say: 3 .... 13 .... 23 28

Once they have firmly fixed the starting and finishing points many children actually close their eyes in order to 'see' the number line more clearly!

A harder example might be: $876 + \boxed{} = 1125$

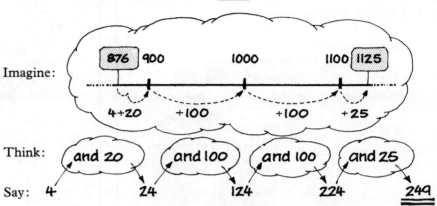

Imagine:

Think:

Say:

This cannot be described as a fully 'mental' process since, in most cases, children either say aloud or jot down the 'running total so far'.

write	think	write
876	900	24
	1100 200	224
	25	249
1125		

Those children who cannot 'hold' figures in their heads will still benefit from attempting the same technique, while jotting down the jumps and running total in the space between starting and finishing numbers.

3 Word Problems

Extending subtraction for four-digit numbers widens the scope for problems. The number of years between historical dates, larger sums of money, lengths, capacities, etc. can now be dealt with. Children should be encouraged to make up or find from within their own experience, problems similar to those provided. This reinforces the ability to 'comb out the mathematics' when making the transition from words to number sentences.

Discussion about the problems and their solution is also very valuable. For example,

'Would the question about the Cortina and the Mini be easier if the Mini's capacity was exactly 1000 cm³?

Is 'making up' easier for this question than 'taking away'?

'Suppose the capacities of both cars is increased by 2 cm³. Will the difference between them be the same?'

Pages from the Pupils' Book and Spiritmasters

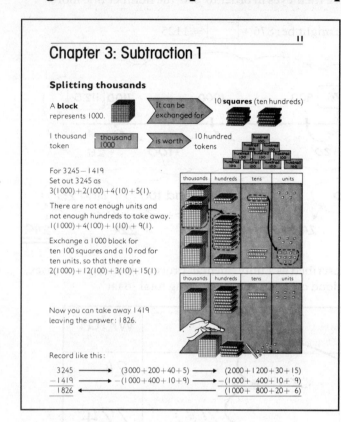

Chapter 3: Subtraction 1

Splitting thousands

A **block** represents 1000. It can be exchanged for 10 **squares** (ten hundreds)

1 thousand token — thousand 1000 — is worth — 10 hundred tokens

For 3245−1419
Set out 3245 as
3(1000)+2(100)+4(10)+5(1).

There are not enough units and not enough hundreds to take away.
1(1000)+4(100)+1(10)+9(1).

Exchange a 1000 block for ten 100 squares and a 10 rod for ten units, so that there are
2(1000)+12(100)+3(10)+15(1).

Now you can take away 1419 leaving the answer: 1826.

Record like this:

3245 →	(3000+200+40+5) →	(2000+1200+30+15)
−1419 →	(1000+400+10+9) →	(1000+ 400+10+ 9)
1826 ←		(1000+ 800+20+ 6)

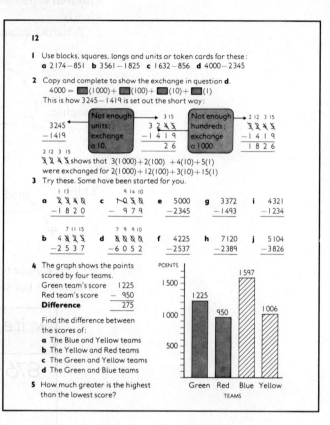

1 Use blocks, squares, longs and units or token cards for these:
   a 2174−851 **b** 3561−1825 **c** 1632−856 **d** 4000−2345

2 Copy and complete to show the exchange in question **d**.
   4000 = ☐(1000)+ ☐(100)+ ☐(10)+ ☐(1)
   This is how 3245−1419 is set out the short way:

   3245 → 3 2̸ 4̸ 5 (Not enough units; exchange a 10.) → 3̸ 2̸ 4̸ 5̸ (Not enough hundreds; exchange a 1000.)
   −1419   −1419   −1419
             2 6   1826

   2̸ 1̸ 3̸ 1̸ 5̸ shows that 3(1000)+2(100) +4(10)+5(1)
   were exchanged for 2(1000)+12(100)+3(10)+15(1).

3 Try these. Some have been started for you.

   a 2̸ 3̸ 4̸ 0̸   **c** 1̸ 0̸ 5̸ 0̸   **e** 5000   **g** 3372   **i** 4321
       −1820   − 979   −2345   −1493   −1234

   b 4 8̸ 2̸ 5̸   **d** 8̸ 0̸ 0̸ 0̸   **f** 4225   **h** 7120   **j** 5104
       −2537   −6052   −2537   −2389   −3826

4 The graph shows the points scored by four teams.
   Green team's score 1225
   Red team's score   − 950
   Difference   275

   Find the difference between the scores of:
   a The Blue and Yellow teams
   b The Yellow and Red teams
   c The Green and Yellow teams
   d The Green and Blue teams

5 How much greater is the highest than the lowest score?

POINTS — Green 1225, Red 950, Blue 1597, Yellow 1006 — TEAMS

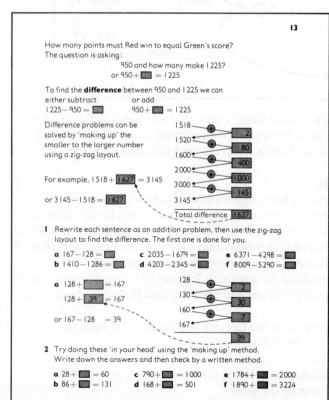

How many points must Red win to equal Green's score?
The question is asking:
                 950 and how many make 1225?
          or 950+ ☐ = 1225

To find the **difference** between 950 and 1225 we can
either subtract   or add
1225−950 = ☐   950+ ☐ = 1225

Difference problems can be solved by 'making up' the smaller to the larger number using a zig-zag layout.

For example, 1518+ 1627 = 3145
or 3145−1518 = 1627

1518	+ 2
1520	+ 80
1600	+ 400
2000	+ 1000
3000	+ 145
3145	
Total difference	1627

1 Rewrite each sentence as an addition problem, then use the zig-zag layout to find the difference. The first one is done for you.

   a 167−128 = ☐   **c** 2035−1679 = ☐   **e** 6371−4298 = ☐
   b 1410−1286 = ☐   **d** 4203−2345 = ☐   **f** 8009−5290 = ☐

   a 128+ ☐ = 167
      128+ 39 = 167
      or 167−128 = 39

128	+ 2
130	+ 30
160	+ 7
167	
	39

2 Try doing these 'in your head' using the 'making up' method. Write down the answers and then check by a written method.

   a 28+ ☐ = 60   **c** 790+ ☐ = 1000   **e** 1784+ ☐ = 2000
   b 86+ ☐ = 131   **d** 168+ ☐ = 501   **f** 1890+ ☐ = 3224

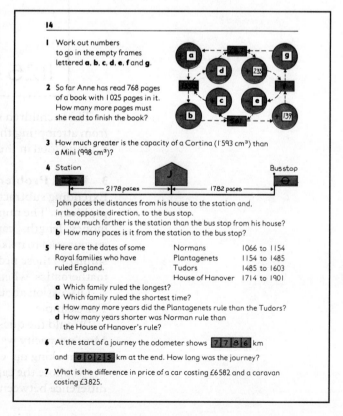

1 Work out numbers to go in the empty frames lettered **a, b, c, d, e, f** and **g**.

2 So far Anne has read 768 pages of a book with 1025 pages in it. How many more pages must she read to finish the book?

3 How much greater is the capacity of a Cortina (1593 cm³) than a Mini (998 cm³)?

4 Station — J — Bus stop
   2178 paces — 1782 paces
   John paces the distances from his house to the station and, in the opposite direction, to the bus stop.
   a How much farther is the station than the bus stop from his house?
   b How many paces is it from the station to the bus stop?

5 Here are the dates of some Royal families who have ruled England.

Normans	1066 to 1154
Plantagenets	1154 to 1485
Tudors	1485 to 1603
House of Hanover	1714 to 1901

   a Which family ruled the longest?
   b Which family ruled the shortest time?
   c How many more years did the Plantagenets rule than the Tudors?
   d How many years shorter was Norman rule than the House of Hanover's rule?

6 At the start of a journey the odometer shows 7 7 8 6 km and 8 0 2 5 km at the end. How long was the journey?

7 What is the difference in price of a car costing £6582 and a caravan costing £3825.

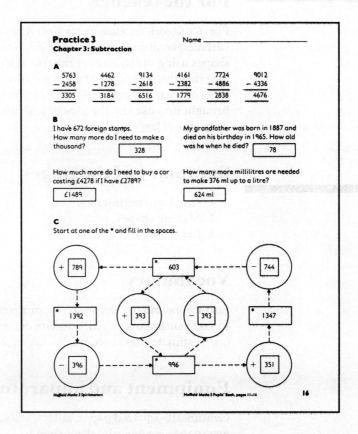

Practice 3
Chapter 3: Subtraction

Name _____

A

5763 − 2458	4462 − 1278	9134 − 2618	4161 − 2382	7724 − 4886	9012 − 4336
3305	3184	6516	1779	2838	4676

B

I have 672 foreign stamps.
How many more do I need to make a
thousand?

328

My grandfather was born in 1887 and
died on his birthday in 1965. How old
was he when he died?

78

How much more do I need to buy a car
costing £4278 if I have £2789?

£1489

How many more millilitres are needed
to make 376 ml up to a litre?

624 ml

C

Start at one of the * and fill in the spaces.

+ 789   603   − 744
1392   + 393   − 393   1347
− 396   996   + 351

Nuffield Maths 5 Spiritmasters   Nuffield Maths 5 Pupils' Book, pages 11–16   16

References and resources

Hart, M. 'Theresa and Subtraction', article in *Mathematics Teaching* No. 87, July 1979

McIntosh, Alistair 'Some Subtractions: What do you think you are doing?' article in *Mathematics Teaching* No. 83, June 1978

Mathematics 5–11 A Handbook of Suggestions, H.M.S.O. 1979

Williams, E. M. and Shuard, H. *Primary Mathematics Today* Third Edition (Chapters 9 and 13), Longman Group Ltd 1982

Arnold, E. J. *Tillich's Base 10 Blocks*

E.S.A. *Base 10 Materials, Multilink cubes*

Metric Aids, (Six to Twelve), *Base 10 Blocks*

Nicolas Burdett *Base 10 Materials*

Taskmaster Ltd, *Base 10 Materials, Decimetric Board*

Chapter 4

Area 1

For the teacher

Previous work on area, in *Nuffield Maths 3* and *4*, has dealt with covering surfaces, counting squares and calculating areas of rectangles and irregular shapes using the square centimetre as a unit. This chapter introduces the idea that the area of a surface is not directly related to its perimeter. Larger units of area, the square decimetre (dm²) and square metre (m²) are brought into use and the idea of scale is introduced in order to draw diagrams of larger areas.

Summary of the stages

1 Areas and perimeters
2 Making shapes
3 Larger areas

Vocabulary

Surface, boundary, perimeter, symmetrical, axis of symmetry, (axes), square centimetre(s) (cm²), square decimetre(s) (dm²), square metre(s) (m²), estimate, row, column, accurate, scale, represents.

Equipment and apparatus

Centimetre-squared paper, rulers, scissors, card, metre rules, string, newspaper, geoboards, elastic bands.

Working with the children

1 Areas and perimeters

The first exercise revises the method of calculating the area of a rectangle by multiplying the number of centimetre squares in a row by the number of rows. Although this amounts to multiplying the number of centimetres in the length by the number of centimetres in the breadth, some children still need to be reminded that in so doing they are really working out the number of centimetre squares required to *cover* the rectangle. The exercise also touches on conservation of area, for although the three rectangles have different dimensions and do not *look* the same, they do have the same area. Some children may need reminding of the 'quick' way to work out the perimeter of a rectangle, $P = 2L + 2B$ or $P = 2(L + B)$; that is:

double the length, double the breadth and add,

or add length to breadth and double the result.
(*Nuffield Maths Spiritmasters 5*, Grid 4)

When the perimeters have been calculated and recorded in the 4th column of the table, the children are asked to write a sentence about the areas and the perimeters. Discussion about the fact that rectangles of the same area can have different perimeters could lead to the possibility of drawing other rectangles of area 24 cm² but with longer perimeters, for example, 24 cm by 1 cm or 48 cm by ½ cm, etc.

Throughout discussion of areas and perimeters it is important to emphasise the *kind* of measurement and hence the units employed:

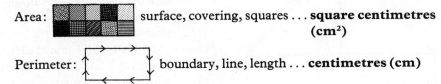

Area: surface, covering, squares ... **square centimetres (cm²)**

Perimeter: boundary, line, length ... **centimetres (cm)**

Further examples of constant area but varying perimeter can be found by using the geoboard.

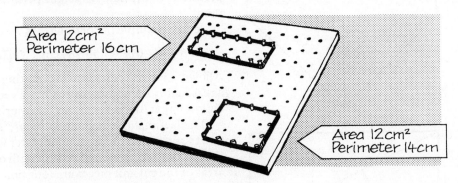

Area 12cm²
Perimeter 16cm

Area 12cm²
Perimeter 14cm

The 'mobility' of the geoboard gives children the opportunity to keep area constant by *putting on* a piece of the same area as that *taken away* – the 'put and take' method.

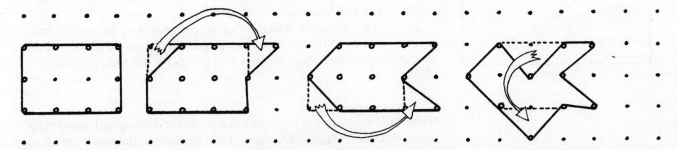

For example, each of these shapes has an area of 6 cm² although the perimeters increase in the diagrams from left to right. This was explained by a boy as, 'I kept the areas the same but I made the perimeters bigger by going in and out more.'

The next exercise reverses the procedure; this time shapes with the same length of boundary or perimeter but with different areas are investigated. (*Nuffield Maths 5 Spiritmasters*, Grid 5.)

Again, some children may be able to pursue the subject further by finding the areas of other rectangles with the same perimeter or by finding which type of rectangle has the maximum area for a fixed perimeter. How should a fence of fixed length be arranged in order to enclose the maximum area of pasture within a straight-sided field?

2 Making shapes

The dissection of a square 10 cm by 10 cm into five pieces – one rectangle and four equal triangles – creates a tangram much simpler than the Chinese

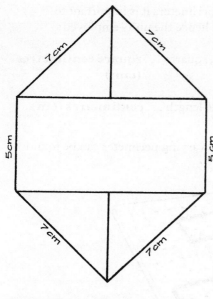

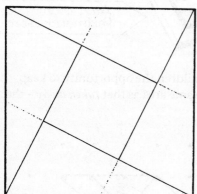

Tangram introduced in *Nuffield Maths 4*. This is deliberate as, although it still gives experience of conservation of area (provided all five pieces are used without overlapping), this simpler example generates more symmetrical shapes than the ancient Chinese Tangram.

The dimensions of the five-piece tangram have also been carefully chosen so that the longest side of each triangle is as close as possible to 7 cm. (In theory it is 7.071 cm.) This makes it possible to measure the perimeters of the shapes made *to nearest centimetre* without too much difficulty. (See Chapter 10 on 'Rounding off'.)

Even if the longest side of each of the triangles forms part of the perimeter of the new shape, the perimeter is still very close to a whole number of centimetres.

In this example the perimeter is very close to 38 cm.

$$5 + 5 + 4(7.071) = 38.284 \text{ cm.}$$ (*Nuffield Maths 5 Spiritmasters*, Grid 27.)

Allowing for errors in drawing, cutting, arranging and measuring, 38 cm is a reasonably accurate result. This is yet another opportunity to remind children that, unlike counting, measuring can never be exact.

The fact that the diagram in *Pupils' Book 5* is drawn to half size also serves as a gentle preparation for later work on scale. Although the sides of the cut-out pieces are twice the length of those shown on the diagram, the areas of the cut-out pieces are four times those of the diagram.

Another interesting dissection of a 10 cm by 10 cm square gives a small square (side approximately 4.5 cm) and four equal right-angled triangles with sides 10 cm, approximately 9 cm and approximately 4.5 cm.

Working round the outer square, each corner in turn is joined to the mid-point of the next side. Unwanted parts of the lines are then erased. (*Nuffield Maths 5 Spiritmasters*, Grid 28.)

In each case the areas of all the shapes created will be the same as the original square (100 cm³) but it is not possible to use all the pieces without overlapping to make a new shape with a perimeter less than the original (40 cm).

3 Larger areas

So far the 1 cm² unit has been quite adequate for dealing with small areas but as children begin to investigate larger surfaces in the environment such as paving stones, window panes, table tops, etc a more convenient size of unit is required. In earlier work on length, the '10 centimetre strip' or *decimetre* was used as a temporary, intermediate unit between the centimetre and the metre. A similar approach is used for area.

The H.M.I. Handbook of Suggestions: Mathematics 5–11 (HMSO, 1979) makes the following comment:

The square decimetre (dm²) is an intermediate measure of surface area which can be used where the square centimetre is too small and the square metre too large.

It is essential that each child makes a square decimetre, preferably of stiff card, linoleum or floor tiling, so that it can be handled and used for comparison. (*Nuffield Maths 5 Spiritmasters*, Grid 20.)

Specific advice on estimation is given because this is an important skill which needs direction and practice. There is little point in paying lip-service to the idea that estimation is a 'good thing' if we do not give guidance and provide opportunities for practice.

The diagrams of the two table tops introduce the idea of *scale* more formally. If possible, the shapes of the two tables should be cut out *to their actual size* from card or sugar paper. This will enable children to compare each scale diagram with the real size it represents and to use square decimetres to 'cover' or to 'step along' the table.

The dimensions of the tables have been chosen so that, although the coffee table has the smaller area, it has the greater perimeter.

Photographs and their enlargements also provide experience of scale allowing children to compare a picture and its 'miniature'.

It is important to encourage the correct language: '1 cm **represents** 10 cm (or 1 dm)' rather than the misleading version, still to be found on some maps or plans, '1 cm = 10 cm', which is an incorrect use of the = sign.

The compound areas A B and C are easily split into rectangles. Drawing the scale versions on cm squared paper gives further experience of using a line 1 cm long to represent 1 dm and a square centimetre to represent a square decimetre. Again, cardboard cut-outs of the actual size will help some children who find it difficult to relate the 'real' to 'scaled down' diagram. (*Nuffield Maths 5 Spiritmasters*, Grid 5.)

Before embarking on the measurement of still larger surfaces using square metres (m²), children should be allowed to 'make a square metre'. If possible a square metre either of paper or enclosed in a string boundary should be left on display so that children have a reference point when discussing areas in square metres.

If four metre rules are used to form a square metre, the corners should be carefully arranged to ensure that the sides of the enclosed area are 1 metre long; that is the rules should not be butted. Osmiroid produce plastic right-angled links for use with Metline metre rules.

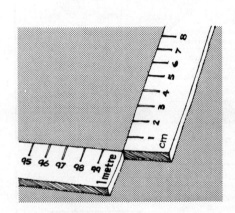

It is helpful for children to have some idea of the area in m² of some familiar rectangles in the environment. A table-tennis table, for example, is approximately 4 square metres.

The following approximate dimensions provide further examples. In order to fit a scale drawing on to A4 size centimetre-squared paper, a suitable scale is also given.

	Length	Breadth	Area	Suggested Scale
Badminton court	13 m	6 m	78 m²	1 cm represents 1 m
Tennis court	24 m	11 m	264 m²	1 cm represents 1 m
Netball court	30 m	15 m	450 m²	1 cm represents 2 m
Full-sized football pitch	100 m	70 m	7000 m²	1 cm represents 5 m

Pages from Pupils' Books and Spiritmasters

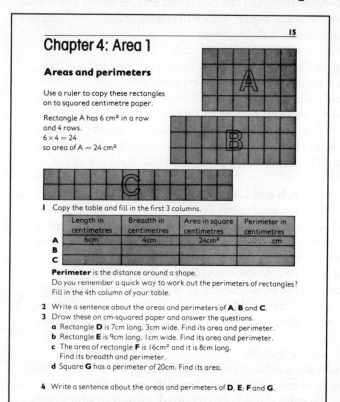

15

Chapter 4: Area 1

Areas and perimeters

Use a ruler to copy these rectangles on to squared centimetre paper.

Rectangle A has 6 cm² in a row and 4 rows.
$6 \times 4 = 24$
so area of A = 24 cm²

1 Copy the table and fill in the first 3 columns.

	Length in centimetres	Breadth in centimetres	Area in square centimetres	Perimeter in centimetres
A	6cm	4cm	24cm²	. . . . .cm
B				
C				

Perimeter is the distance around a shape.
Do you remember a quick way to work out the perimeters of rectangles?
Fill in the 4th column of your table.

2 Write a sentence about the areas and perimeters of **A**, **B** and **C**.
3 Draw these on cm-squared paper and answer the questions.
   a Rectangle **D** is 7cm long, 3cm wide. Find its area and perimeter.
   b Rectangle **E** is 9cm long, 1cm wide. Find its area and perimeter.
   c The area of rectangle **F** is 16cm² and it is 8cm long.
     Find its breadth and perimeter.
   d Square **G** has a perimeter of 20cm. Find its area.

4 Write a sentence about the areas and perimeters of **D**, **E**, **F** and **G**.

16

Making shapes

Cut out a square 10cm by 10cm from card.

Mark and cut out the 5 shapes: 1 rectangle and 4 equal triangles. (The diagrams are half size.)

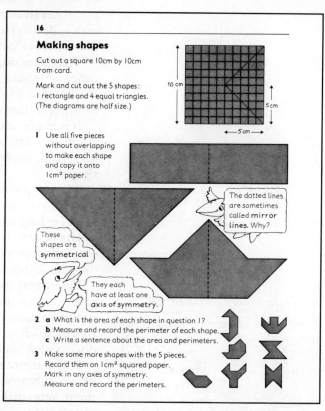

1 Use all five pieces without overlapping to make each shape and copy it onto 1cm² paper.

These shapes are symmetrical

They each have at least one axis of symmetry.

The dotted lines are sometimes called mirror lines. Why?

2 a What is the area of each shape in question 1?
  b Measure and record the perimeter of each shape.
  c Write a sentence about the area and perimeters.

3 Make some more shapes with the 5 pieces.
  Record them on 1cm² squared paper.
  Mark in any axes of symmetry.
  Measure and record the perimeters.

17

Larger areas

On cm² paper draw a square with sides 10cm long.

Carefully cut out the square—you will need it later.

Copy the sentence about the area on to your square.

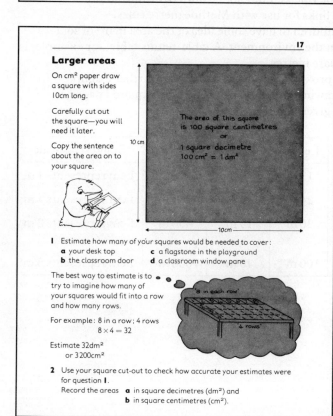

The area of this square is 100 square centimetres
or
1 square decimetre
100 cm² = 1 dm²

1 Estimate how many of your squares would be needed to cover:
  a your desk top
  b the classroom door
  c a flagstone in the playground
  d a classroom window pane

The best way to estimate is to try to imagine how many of your squares would fit into a row and how many rows.

For example: 8 in a row; 4 rows
$8 \times 4 = 32$

Estimate 32dm²
   or 3 200cm²

2 Use your square cut-out to check how accurate your estimates were for question 1.
  Record the areas a in square decimetres (dm²) and
                    b in square centimetres (cm²).

18

The diagrams of two table tops had to be drawn smaller than they really are in order to fit them on this page.

They are drawn to **scale**.

1cm represents 10cm (or 1dm)
1cm² represents 100cm² (or 1dm²)

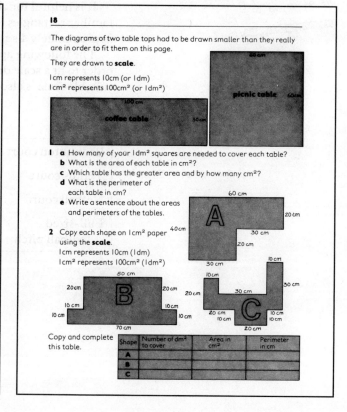

1 a How many of your 1dm² squares are needed to cover each table?
  b What is the area of each table in cm²?
  c Which table has the greater area and by how many cm²?
  d What is the perimeter of each table in cm²?
  e Write a sentence about the areas and perimeters of the tables.

2 Copy each shape on 1cm² paper using the **scale**.
  1cm represents 10cm (1dm)
  1cm² represents 100cm² (1dm²)

Copy and complete this table.

Shape	Number of dm² to cover	Area in cm²	Perimeter in cm
A			
B			
C			

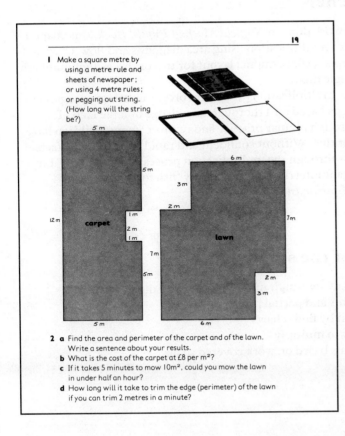

19

I Make a square metre by using a metre rule and sheets of newspaper; or using 4 metre rules; or pegging out string. (How long will the string be?)

carpet

lawn

2 **a** Find the area and perimeter of the carpet and of the lawn. Write a sentence about your results.
  b What is the cost of the carpet at £8 per m²?
  c If it takes 5 minutes to mow 10m², could you mow the lawn in under half an hour?
  d How long will it take to trim the edge (perimeter) of the lawn if you can trim 2 metres in a minute?

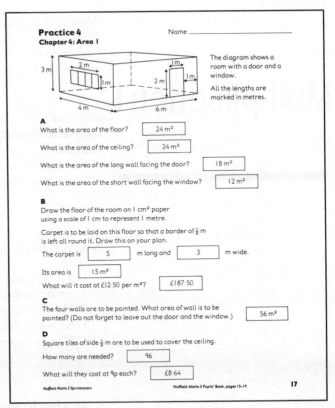

Practice 4
Chapter 4: Area 1

Name _____

The diagram shows a room with a door and a window.

All the lengths are marked in metres.

A
What is the area of the floor? `24 m²`

What is the area of the ceiling? `24 m²`

What is the area of the long wall facing the door? `18 m²`

What is the area of the short wall facing the window? `12 m²`

B
Draw the floor of the room on I cm² paper using a scale of I cm to represent I metre.

Carpet is to be laid on this floor so that a border of ½ m is left all round it. Draw this on your plan.

The carpet is `5` m long and `3` m wide.

Its area is `15 m²`

What will it cost at £12·50 per m²? `£187·50`

C
The four walls are to be painted. What area of wall is to be painted? (Do not forget to leave out the door and the window.) `56 m²`

D
Square tiles of side ½ m are to be used to cover the ceiling.

How many are needed? `96`

What will they cost at 9p each? `£8·64`

Nuffield Maths 5 Spiritmasters Nuffield Maths 5 Pupils' Book, pages 15–19 **17**

References and resources

Nuffield Mathematics Teaching Project, *Shape and Size* ▽, *Shape and Size* ▽ Nuffield Guides, Chambers/Murray 1967 (See Introduction, page xi.)

Williams, E. M. and Shuard, H. *Primary Mathematics Today* Third Edition (Chapters 7 and 23), Longman Group Ltd 1982

E. J. Arnold *Geoboards*

Invicta Plastics, *Area measuring grid, Centimetre pin board*

Osmiroid, *Metline metre rules, Right-angled links*

Philip & Tacey Ltd, *Coloured gummed area grid paper, N.E.T. Square centimetre printed cards, Plastic area grids, Square metre tiles (4¼-metre tiles)*

Taskmaster Ltd, *Centimetre grid*

Triman Classmate, *Cover-up*

Multiplication 1

For the teacher

After revising the work done in *Nuffield Maths 4 Pupils' Book*, this chapter extends the scope of multiplication using area diagrams and flow-charts, leading to the compact, conventional layout for two or three-digit numbers multiplied by a single digit.

At this level, the multiplication process involves the stringing together of several small steps based on a thorough knowledge of table facts. It is important that, through regular practice and reinforcement, children have these at their finger tips. Without confident and rapid recall of table facts, children's progress through the multiplication process will not 'flow' but be subject to constant interruptions while they 'fish for a product'. (*Nuffield Maths 5 Spiritmasters*, Grid 22.)

Summary of the stages

1 Multiplying any two-digit number by a single digit (using diagrams and partial products)
2 Multiplication by flow-charts
3 A shorter way to multiply
4 Word problems based on work covered so far

Vocabulary

Digit, product, partial product, operation.

Equipment and apparatus

Squared paper, graph paper (with small squares), rulers, base 10 materials.

Working with the children

1 Multiplying any two-digit number by a single digit

In *Pupils' Book 4* the distributive property of multiplication over addition allowed us to separate one multiplication into two simpler computations.

For example, 17×6 could be written as:

$$
\begin{array}{l}
10 + 7 \\
\underline{\times6} \\
60 + 42 = \underline{102}
\end{array}
\qquad
\begin{aligned}
\text{or} \quad 17 \times 6 &= 10 \times 6 + 7 \times 6 \\
&= 60 + 42 \\
&= \underline{102}
\end{aligned}
$$

In *Pupils' Book 5* an area diagram is linked with a partial product layout at the side. The units are multiplied first as this will be the order of multiplication when the shorter layout is introduced at a later stage.

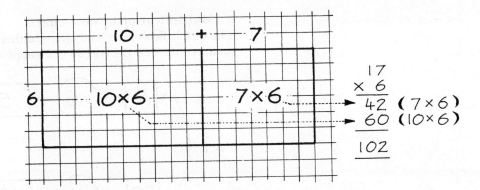

It is important to remind the children that careful presentation, with figures in the correct columns, is essential.

So far the children have been able to draw their diagrams on centimetre-squared paper but this will no longer be possible with numbers greater than 20. To continue the area diagram approach, graph paper based on a smaller unit is required. Metric graph paper which is divided into 2 cm squares and then sub-divided into 2 mm squares is suitable but any graph paper with a reasonably small unit will do.

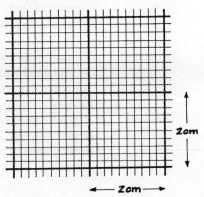

As before, the area diagram is linked to the partial product layout at the side:

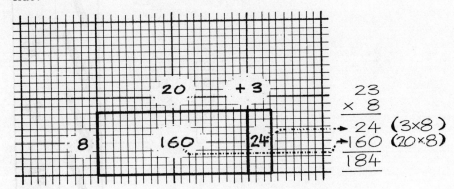

Again use is made of the distributive property of multiplication over addition so that 23×8 becomes $(20 + 3) \times 8$.

Some children will be ready before others to complete the multiplication without a diagram; the timing of this step is left to the discretion of the teacher.

2 Multiplication by flow-charts

The simple flow-chart is suggested either as an alternative or as reinforcement for the area diagram approach.

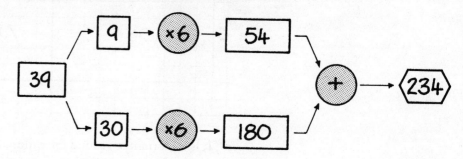

In this example 39 is written in the input frame and then expanded to 30 and 9. The operation is to multiply by 6 so ' × 6' is written in the operation frames. Carrying out this multiplication gives the partial products which are then added together to give the answer.

Squares are used for **input** numbers, circles for **operations**, oblongs for **partial products** and hexagons for the **answer**. To save time, it is suggested that the children use prepared layouts. (*Nuffield Maths 5 Spiritmasters*, Grid 19.)

3 A shorter way to multiply

In order to introduce the shorter, conventional layout, a brief return is made to base ten apparatus in order to demonstrate what the 'on-paper' calculation represents. As a preliminary, it is as well to remind children that multiplication is repeated addition. The picture of apparatus in *Pupils' Book 5* is really a shortened version of 'Set out 1 ten-rod and 3 units four times and then put them together.'

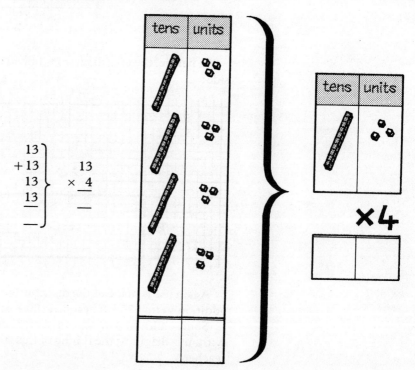

In effect the shorter layout merely involves 'telescoping' two of the steps of the partial product method into one.

For example: 26×3

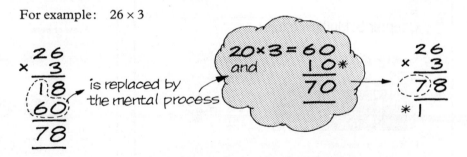

The 1 * in the tens column below the answer box serves as a reminder of the 1 ten-rod waiting to be gathered during the 'apparatus' version.

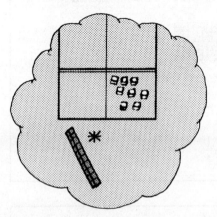

Even when children show confidence in using the shorter layout without reference to apparatus, it is a good idea to ask them occasionally to use base ten pieces or tokens to demonstrate what the 'on-paper' calculation represents.

4   Word problems based on work covered so far

Once again it cannot be stressed too greatly how important word problems are. Children need to appreciate how calculation is used to deal with problems that arise in everyday life. They should also be encouraged to write their own problems or stories whenever they are dealing with abstract number.

Pages from the Pupils' Books and Spiritmasters

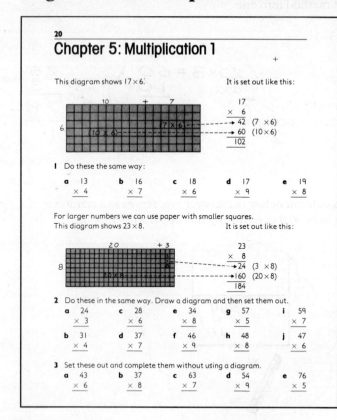

20

Chapter 5: Multiplication 1

This diagram shows 17 × 6.   It is set out like this:

```
         17
       ×  6
    →   42   (7 × 6)
    →   60   (10 × 6)
        102
```

1 Do these the same way:

a 13	**b** 16	**c** 18	**d** 17	**e** 19
× 4	× 7	× 6	× 9	× 8

For larger numbers we can use paper with smaller squares.
This diagram shows 23 × 8.   It is set out like this:

```
         23
       ×  8
    →   24   (3 × 8)
    →  160   (20 × 8)
       184
```

2 Do these in the same way. Draw a diagram and then set them out.

a 24	**c** 28	**e** 34	**g** 57	**i** 59
× 3	× 6	× 8	× 5	× 7
b 31	**d** 37	**f** 46	**h** 48	**j** 47
× 4	× 7	× 9	× 8	× 6

3 Set these out and complete them without using a diagram.

a 43	**b** 37	**c** 63	**d** 54	**e** 76
× 6	× 8	× 7	× 9	× 5

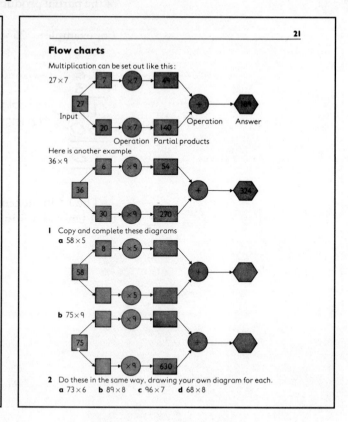

21

Flow charts

Multiplication can be set out like this:

27 × 7

Input — Operation — Partial products — Operation — Answer

Here is another example
36 × 9

1 Copy and complete these diagrams
a 58 × 5

b 75 × 9

2 Do these in the same way, drawing your own diagram for each.
a 73 × 6   **b** 89 × 8   **c** 96 × 7   **d** 68 × 8

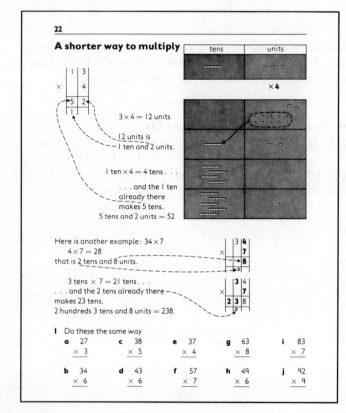

22

A shorter way to multiply

	tens	units
		× 4

3 × 4 = 12 units

12 units is
1 ten and 2 units.

1 ten × 4 = 4 tens . . .

. . . and the 1 ten
already there
makes 5 tens.
5 tens and 2 units = 52

Here is another example: 34 × 7
4 × 7 = 28
that is 2 tens and 8 units.

3 tens × 7 = 21 tens . . .
. . . and the 2 tens already there
makes 23 tens.
2 hundreds 3 tens and 8 units = 238.

1 Do these the same way

a 27	**c** 38	**e** 37	**g** 63	**i** 83
× 3	× 5	× 4	× 8	× 7
b 34	**d** 43	**f** 57	**h** 49	**j** 92
× 6	× 6	× 7	× 6	× 9

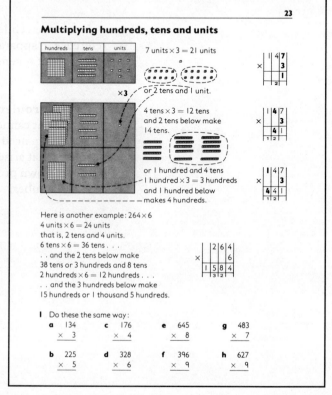

23

Multiplying hundreds, tens and units

hundreds	tens	units
		× 3

7 units × 3 = 21 units
or 2 tens and 1 unit.

4 tens × 3 = 12 tens
and 2 tens below make
14 tens.

or 1 hundred and 4 tens

1 hundred × 3 = 3 hundreds
and 1 hundred below
makes 4 hundreds.

Here is another example : 264 × 6
4 units × 6 = 24 units
that is, 2 tens and 4 units.
6 tens × 6 = 36 tens . . .
. . and the 2 tens below make
38 tens or 3 hundreds and 8 tens
2 hundreds × 6 = 12 hundreds . . .
. . and the 3 hundreds below make
15 hundreds or 1 thousand 5 hundreds.

1 Do these the same way :

a 134	**c** 176	**e** 645	**g** 483
× 3	× 4	× 8	× 7
b 225	**d** 328	**f** 396	**h** 627
× 5	× 6	× 9	× 9

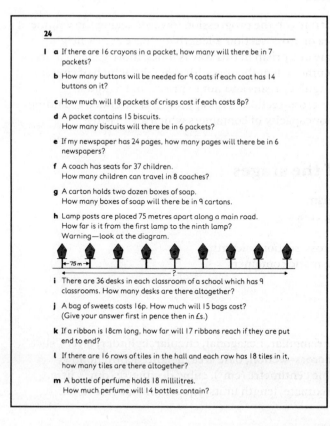

24

1 a If there are 16 crayons in a packet, how many will there be in 7 packets?

b How many buttons will be needed for 9 coats if each coat has 14 buttons on it?

c How much will 18 packets of crisps cost if each costs 8p?

d A packet contains 15 biscuits.
How many biscuits will there be in 6 packets?

e If my newspaper has 24 pages, how many pages will there be in 6 newspapers?

f A coach has seats for 37 children.
How many children can travel in 8 coaches?

g A carton holds two dozen boxes of soap.
How many boxes of soap will there be in 9 cartons?

h Lamp posts are placed 75 metres apart along a main road.
How far is it from the first lamp to the ninth lamp?
Warning—look at the diagram.

i There are 36 desks in each classroom of a school which has 9 classrooms. How many desks are there altogether?

j A bag of sweets costs 16p. How much will 15 bags cost?
(Give your answer first in pence then in £s.)

k If a ribbon is 18cm long, how far will 17 ribbons reach if they are put end to end?

l If there are 16 rows of tiles in the hall and each row has 18 tiles in it, how many tiles are there altogether?

m A bottle of perfume holds 18 millilitres.
How much perfume will 14 bottles contain?

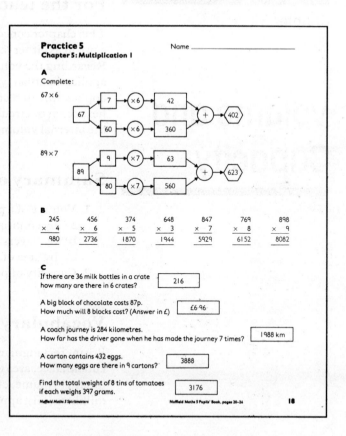

Practice 5
Chapter 5: Multiplication 1

Name _____

A
Complete:

67 × 6

89 × 7

B

245	456	374	648	847	769	898
× 4	× 6	× 5	× 3	× 7	× 8	× 9
980	2736	1870	1944	5929	6152	8082

C

If there are 36 milk bottles in a crate how many are there in 6 crates? ☐ 216

A big block of chocolate costs 87p.
How much will 8 blocks cost? (Answer in £) ☐ £6·96

A coach journey is 284 kilometres.
How far has the driver gone when he has made the journey 7 times? ☐ 1988 km

A carton contains 432 eggs.
How many eggs are there in 9 cartons? ☐ 3888

Find the total weight of 8 tins of tomatoes if each weighs 397 grams. ☐ 3176

Nuffield Maths 5 Spiritmasters   *Nuffield Maths 5 Pupils' Book, pages 20–24*   **18**

References and resources

Cuff, C. *Tables Trail*, Longman Group Ltd 1978

Nuffield Mathematics Teaching Project, *Computation and Structure* ③ Nuffield Guide, Chambers/Murray 1967 (See Introduction, page xi.)

Williams, E. M. and Shuard, H. *Primary Mathematics Today* Third Edition (Chapters 15 and 17), Longman Group Ltd 1982

Volume and Capacity

For the teacher

This chapter concentrates on the progression towards seeing the volume of a prism as either 'area of cross-section × length' or 'base area × height'. Regarding the volume of a prism in this way is much more general in its application than 'volume = length × breadth × height' which of course applies only to rectangular prisms and not to prisms of triangular, hexagonal or circular cross-section. This approach is extended to finding the internal volume or capacity of containers which are prisms.

Summary of the stages

1 Volume of a prism:
   a) by counting cubes
   b) by 'layers'
   c) by area of cross-section × length
2 Capacity of prismatic containers

Vocabulary

Prism, rectangular, triangular, hexagonal, circular, cylinder, layer ('slice'), cuboid, face, area of cross-section, base area, height, depth, hollow, internal volume, cubic centimetre (cm³), cubic decimetre (dm³), litre, millilitre (ml), approximate, length units, square units, cubic units.

Equipment and apparatus

Centimetre cubes, solid shapes, hollow boxes or containers (some prisms, some non-prisms), centimetre-squared paper.

Working with the children

1 a) Volume of a prism by counting cubes
Rather than being given a formal definition involving such expressions as 'solid of uniform cross-section' or 'congruent parallel faces', the children are encouraged to think of a prism as a solid shape which, if it were made of bread, could be cut into 'slices' all the same shape and size. Perhaps it is as well to emphasise that we are referring to 'slices' of a cut loaf rather than the 'wedges' cut from a melon or spherical cheese!

The first exercise, designed to ensure that children can differentiate between prisms and non-prisms, should be supplemented by boxes, containers and shapes made from wood, plastic or modelling clay.

Earlier activities with cubes (*Nuffield Maths 2 Teachers' Handbook*, page 145; *Nuffield Maths 3 Pupils' Book* page 112; *Nuffield Maths 4 Pupils' Book* pages 31 and 33) may have provided some children with sufficient experience to interpret diagrams in which not all the cubes can be seen. However, it is much better if the retangular prisms are *actually built* so that the cubes used can be physically counted rather than 'imagined'. This activity will lead naturally to the situation where the children can be asked 'Do we need to count every cube or is there a quicker way?'

1 b) Volume of a prism by 'layers'

The technique for finding the volume of a rectangular prism by finding how many cubes in a 'layer' or 'slice' and multiplying by the number of layers was referred to in Chapter 8 of *Nuffield Maths 4 Pupils' Book*. If coloured cubes are available this can be reinforced by using colours to emphasise the 'slice' or 'layer' at one end and the separate 'slices' or 'layers'.

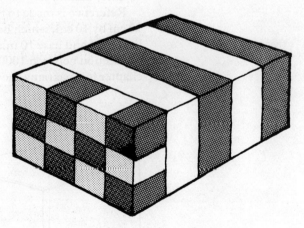

1 c) Volume of prism by area of cross-section × length

The next step is for the children to appreciate that counting the number of cubes in the end 'slice' is like finding the area of the end face of the prism.

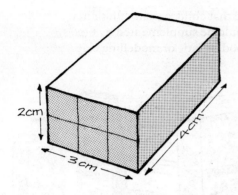

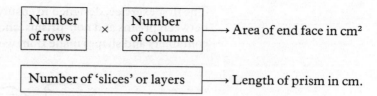

→ Area of end face in cm²

→ Length of prism in cm.

'Area of cross-section' is introduced in place of 'slice area' to give the formula:

$$\text{Volume of a cuboid} = \text{area of cross-section} \times \text{length}$$

(*Nuffield Maths 5 Spiritmasters*, Grid 14 and 15.)

After an exercise involving rectangular prisms or cuboids only, this is extended to include all prisms. At this point it is important to stress conformity of units – if the area of cross-section is measured in square centimetres (cm²) and the length in centimetres (cm), then the volume will be in cubic centimetres (cm³). Obviously, the volume of a cuboid of wood 1 metre long and with a cross-section of 1 cm² cannot be found by multiplying 1 cm² by 1 m! All the dimensions must be measured using the same basic unit: $1 \text{ cm}^2 \times 100 \text{ cm} = 100 \text{ cm}^3$.

2 Capacity of prismatic containers

When finding the internal volume or capacity of containers in the shape of prisms, the base area is chosen as the area of cross-section. In cases where the base is not a rectangle (cylinders, triangular or hexagonal prisms, for example) the 'counting centimetre squares' technique is used to find the approximate base area. The length of the prism, now referred to as the height, is measured to the nearest centimetre or half-centimetre. (See Chapter 10 on 'Rounding off'.) The approximate capacity worked out using this type of calculation can be compared with the result achieved by filling the container with water or fine sand and pouring into a graduated capacity measure.

Reference is made to the hollow *decimetre cube* measuring 10 cm by 10 cm by 10 cm, which has a capacity of 1000 cm³ or 1 *litre*. A hollow plastic litre cube and base 10 m.a.b. material are very useful for demonstrating the connection between 1000 cm³, 1000 ml and the litre. (See Appendix of this chapter for diagrams showing how to make a litre cube from stiff paper.)

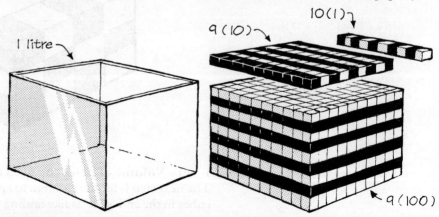

Pages from the Pupils' Book and Spiritmasters

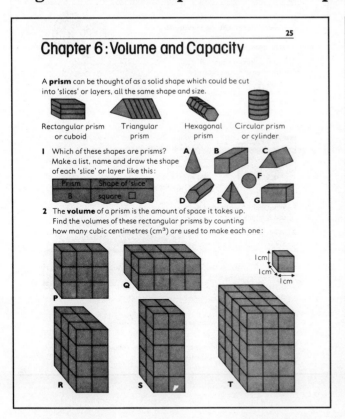

Chapter 6 : Volume and Capacity

A **prism** can be thought of as a solid shape which could be cut into 'slices' or layers, all the same shape and size.

Rectangular prism or cuboid   Triangular prism   Hexagonal prism   Circular prism or cylinder

1 Which of these shapes are prisms? Make a list, name and draw the shape of each 'slice' or layer like this :

Prism	Shape of 'slice'
B	square □

2 The **volume** of a prism is the amount of space it takes up. Find the volumes of these rectangular prisms by counting how many cubic centimetres (cm³) are used to make each one :

P   Q   R   S   T

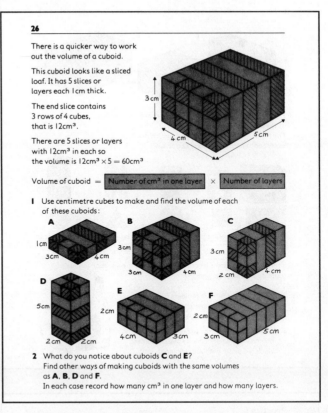

There is a quicker way to work out the volume of a cuboid.

This cuboid looks like a sliced loaf. It has 5 slices or layers each 1cm thick.

The end slice contains 3 rows of 4 cubes, that is 12cm³.

There are 5 slices or layers with 12cm³ in each so the volume is 12cm³ × 5 = 60cm³

Volume of cuboid = Number of cm³ in one layer × Number of layers

1 Use centimetre cubes to make and find the volume of each of these cuboids :

A   B   C   D   E   F

2 What do you notice about cuboids **C** and **E**? Find other ways of making cuboids with the same volumes as **A**, **B**, **D** and **F**. In each case record how many cm³ in one layer and how many layers.

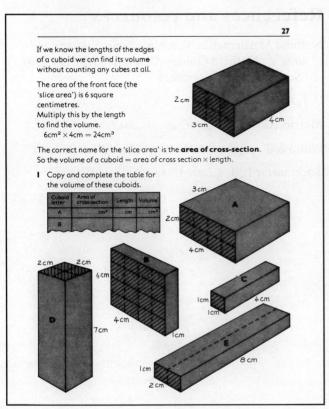

If we know the lengths of the edges of a cuboid we can find its volume without counting any cubes at all.

The area of the front face (the 'slice area') is 6 square centimetres. Multiply this by the length to find the volume.
6cm² × 4cm = 24cm³

The correct name for the 'slice area' is the **area of cross-section**. So the volume of a cuboid = area of cross section × length.

1 Copy and complete the table for the volume of these cuboids.

Cuboid letter	Area of cross-section	Length	Volume
A	cm²	cm	cm³
B			

A   B   C   D   E

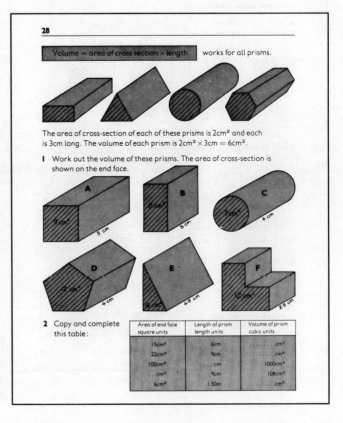

Volume = area of cross section × length   works for all prisms.

The area of cross-section of each of these prisms is 2cm² and each is 3cm long. The volume of each prism is 2cm² × 3cm = 6cm³.

1 Work out the volume of these prisms. The area of cross-section is shown on the end face.

A   B   C   D   E   F

2 Copy and complete this table:

Area of end face square units	Length of prism length units	Volume of prism cubic units
15cm²	6cm	......cm³
22cm²	9cm	......cm³
100cm²	.....cm	1000cm³
.....cm²	9cm	108cm³
6cm²	1·50m	......cm³

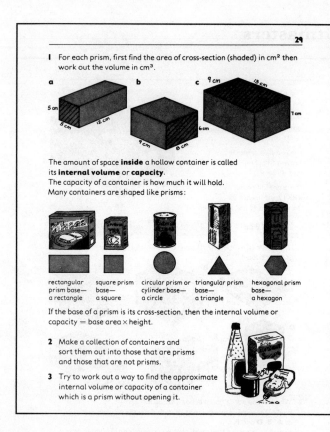

29

1   For each prism, first find the area of cross-section (shaded) in cm² then work out the volume in cm³.

a 5 cm, 5 cm, 12 cm

b 9 cm, 6 cm, 8 cm

c 9 cm, 13 cm, 7 cm

The amount of space **inside** a hollow container is called its **internal volume** or **capacity**.
The capacity of a container is how much it will hold.
Many containers are shaped like prisms:

rectangular prism base— a rectangle	square prism base— a square	circular prism or cylinder base— a circle	triangular prism base— a triangle	hexagonal prism base— a hexagon

If the base of a prism is its cross-section, then the internal volume or capacity = base area × height.

2   Make a collection of containers and sort them out into those that are prisms and those that are not prisms.

3   Try to work out a way to find the approximate internal volume or capacity of a container which is a prism without opening it.

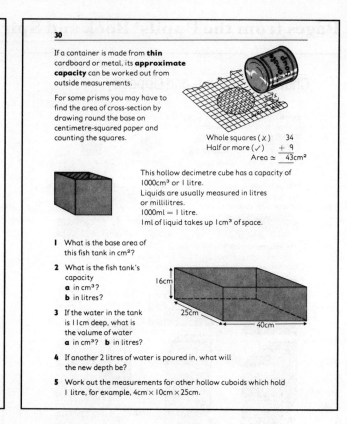

30

If a container is made from **thin** cardboard or metal, its **approximate capacity** can be worked out from outside measurements.

For some prisms you may have to find the area of cross-section by drawing round the base on centimetre-squared paper and counting the squares.

Whole squares (x)   34
Half or more (✓)   + 9
Area ≃   43cm²

This hollow decimetre cube has a capacity of 1000cm³ or 1 litre.
Liquids are usually measured in litres or millilitres.
1000ml = 1 litre.
1ml of liquid takes up 1cm³ of space.

1   What is the base area of this fish tank in cm²?

2   What is the fish tank's capacity
    a in cm³?
    b in litres?

3   If the water in the tank is 11cm deep, what is the volume of water
    a in cm³?   **b** in litres?

4   If another 2 litres of water is poured in, what will the new depth be?

5   Work out the measurements for other hollow cuboids which hold 1 litre, for example, 4cm × 10cm × 25cm.

16cm, 25cm, 40cm

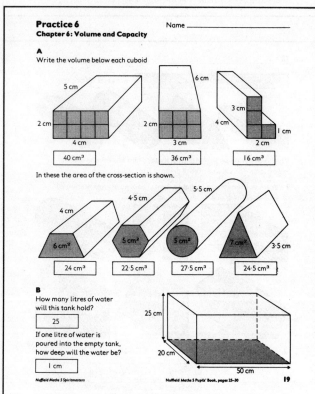

Practice 6
Chapter 6: Volume and Capacity

Name _____

A
Write the volume below each cuboid

5 cm, 2 cm, 4 cm   40 cm³

6 cm, 2 cm, 3 cm   36 cm³

3 cm, 4 cm, 2 cm, 1 cm   16 cm³

In these the area of the cross-section is shown.

4 cm, 6 cm²   24 cm³

4·5 cm, 5 cm²   22·5 cm³

5·5 cm, 5 cm²   27·5 cm³

7 cm², 3·5 cm   24·5 cm³

B
How many litres of water will this tank hold?

25

If one litre of water is poured into the empty tank, how deep will the water be?

1 cm

25 cm, 20 cm, 50 cm

Nuffield Maths 5 Spiritmasters   *Nuffield Maths 5 Pupils' Book, pages 25–30*   **19**

References and resources

Nuffield Mathematics Teaching Project, *Shape and Size* ▽, Nuffield Guides, Chambers/Murray 1967 (See Introduction, page xi.)

E. J. Arnold *Litre Cube*

Metric-Aids, *Capacity Cubes, Cubic Litre*

Philip & Tacey Ltd, *Cubic Centimetre Plastic Cubes*

Taskmaster Ltd, *1 Litre Cube*

Appendix

Construction of 1 litre container from stiff paper, without glue.

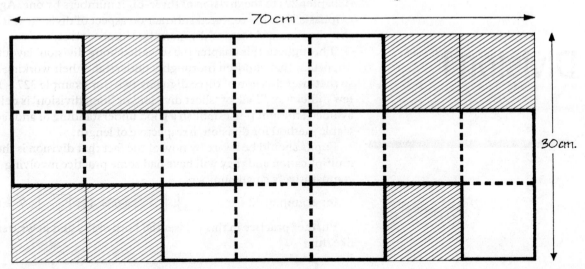

Net for 1 litre open box. Squares 10cm x 10cm.

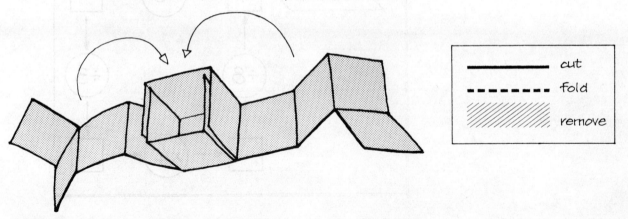

——————	cut
- - - - - -	fold
▨▨▨	remove

Division 1

For the teacher

This chapter extends the work of *Nuffield Maths 4 Teachers' Handbook* (Chapter 22) to the division of three-digit numbers by one. Again the emphasis is on the repeated subtraction aspect of division as it is from this that the standard form of division algorithm evolves.

Throughout this chapter the so-called 'long division' layout is used. It is important that children thoroughly understand their working at this stage so that later division of three digits by two (for example $327 \div 17$) presents few difficulties. Talk of 'short division' or 'long division' is deliberately avoided; the aim is to establish a good understanding of and facility with a viable method for division, irrespective of length.

Pupils should be aware by now of the fact that division is the inverse of multiplication and they will have had some practice involving comparatively small numbers

for example: $42 \div 6 = \boxed{}$; $\boxed{} \times 6 = 42$; $\boxed{} \div 7 = 6$

Further practice in this is essential both orally and in written problems like this:

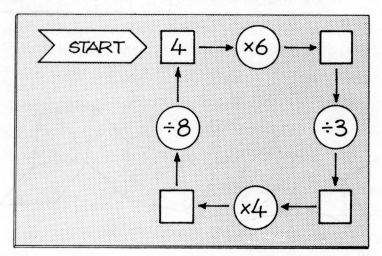

These flow diagrams are not difficult to manufacture and can be simple or complicated as the need arises. (*Nuffield Maths 5 Spiritmasters*, Grid 21.)

Summary of the stages

1 Revision of previous work leading to division of two digits by one with answers greater than 20
2 Division of three digits by one with answers less than 100
3 Revision of multiplication by 10 and multiples of 10 leading to a shorter layout

Vocabulary

Division, subtract, take away, multiple, digit.

Equipment and apparatus

Number line, multiplication square.

Working with the children

1 Revision of previous work leading to division of two digits by one with answers greater than 20

After brief revision of the work covered in *Pupils' Book 4*, the work is extended to give answers greater than 20. The repeated subtraction approach may appear 'long-winded' but it is understandable to the child and forms the basis for more difficult work.

The 'slick trick' or 'rote rule' may enable the pupils to do these simple divisions but division by two digits will be a different and, for too many, a painful experience! It must be pointed out too that many of the 'traditional' rituals are mathematically incorrect. Take the example:

$$3 \overline{)73}$$

Traditionally pupils have been taught to begin '3 into 7 goes . . . .' Is it 7? Haven't we just spent 4 or 5 years showing and discovering that the 7 represents 70?

The suggested layout removes the need for this incorrect language and concentrates on subtracting '10 lots of 3' as many times as possible.

$$
\begin{array}{r}
3 \overline{)73} \\
-30 \quad 10\,(3) \\
\hline
43
\end{array}
$$

10 lots of 3 are taken away and 43 is left.
Another 10 lots of 3 can be taken away.

$$
\begin{array}{r}
3 \overline{)73} \\
-30 \;|\; 10\,(3) \\
\hline
43 \\
-30 \;|\; 10\,(3) \\
\hline
13
\end{array}
$$

From the 13 left, 4 lots of 3 can be taken.

$$
\begin{array}{r}
24 \; r \; 1 \\
3 \overline{)73} \\
-30 \;|\; 10\,(3) \\
\hline
43 \\
-30 \;|\; 10\,(3) \\
\hline
13 \\
-12 \;|\; 4\,(3) \\
\hline
1 \;|\; 24\,(3)
\end{array}
$$

24 lots of 3 altogether.

In this chapter problems with, and without, remainders are mixed as by now the pupil should be handling remainders without difficulty.

2 Division of three digits by one with answers less than 100

The examples on this page are similar in layout to the one above but with three figure dividends.

Care has been taken in these examples to keep the number of subtractions to a maximum of four as this stage needs to go on to the next as soon as possible.

3 Revision of multiplication by 10 and multiples of 10 leading to a shorter layout

In their work on place value, the pupils will have multiplied by 10 by moving units to the tens column and filling in the units space with a zero. We need to revise this before taking the next step in division. It cannot be over-emphasised that multiplication by 10 using the rule 'to multiply by 10 add a nought' should not be taught. Does $7.3 \times 10 = 7.30$? Of course not!

Column boards may be used to reinforce the method.

The columns are marked on the board and digits on cards represent the numbers. Here the physical picking up and moving of the digits underlines the concept.

Alternatively, multiplication by 10 can be shown on an abacus:

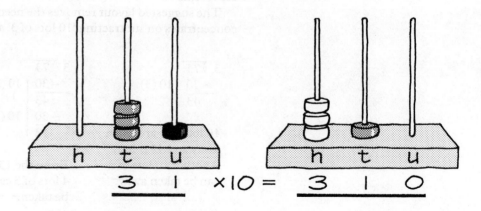

$$31 \times 10 = 310$$

or by using base ten blocks on a counting board:

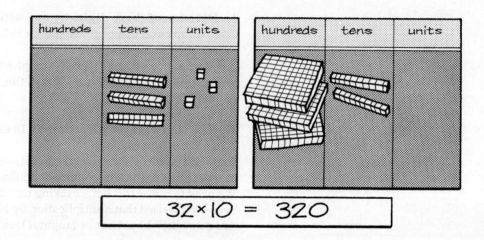

$$32 \times 10 = 320$$

Another method of demonstrating the multiplication by 10 can be made from card:

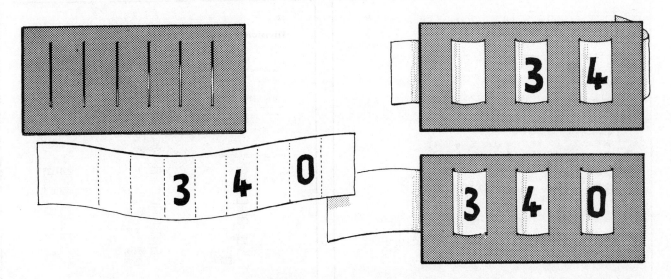

Six slits are cut in the card and strips made with spaces matching those between the slits. The strip can be threaded onto the card so that the first two digits show in the right-hand windows. To multiply by 10, we pull the strip along and a zero fills the gap in the units.

The multiplication by multiples of ten is carried out in two steps:

$$27 \times 40 = \quad 27 \times 4 \times 10$$

Multiply by the units $\quad\quad = \quad 108 \times 10$

Multiply by ten $\quad\quad\quad = 1080$

Having revised this, the layout of division can now be shortened. Our aim should be to use the shortest possible layout but some pupils will not achieve this aim quickly. Children should be encouraged to try to reduce the length of the layout – but not at the expense of accuracy.

```
        36 r 3                        36 r 3
    9 )327                        9 )327
     −270 | 30 (9)                 −180 | 20 (9)
       57 |                         147 |
     − 54 | 6 (9)                  − 90 | 10 (9)
        3 | 36 (9)                   57 |
                                   − 45 | 5 (9)
                                     12 |
                                   −  9 | 1 (9)
                                      3 | 36 (9)
```

Whilst this layout
should be the aim,
this is equally correct.

41

Pages from the Pupils' Book and Spiritmasters

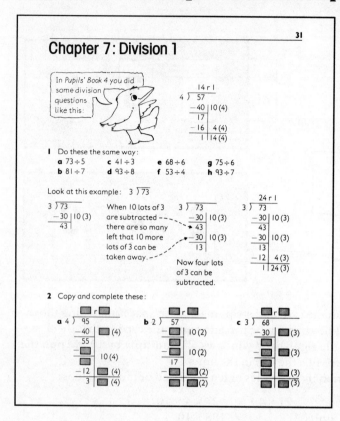

31

Chapter 7: Division 1

In Pupils' Book 4 you did some division questions like this:

```
        14 r 1
    4 ) 57
      - 40  | 10 (4)
        17
      - 16  |  4 (4)
         1  | 14 (4)
```

1 Do these the same way:
   a 73 ÷ 5   c 41 ÷ 3   e 68 ÷ 6   g 75 ÷ 6
   b 81 ÷ 7   d 93 ÷ 8   f 53 ÷ 4   h 93 ÷ 7

Look at this example: 3) 73

```
  3 ) 73
    - 30 | 10 (3)
      43
```
When 10 lots of 3 are subtracted there are so many left that 10 more lots of 3 can be taken away.

```
  3 ) 73
    - 30 | 10 (3)
      43
    - 30 | 10 (3)
      13
```
Now four lots of 3 can be subtracted.

```
        24 r 1
  3 ) 73
    - 30 | 10 (3)
      43
    - 30 | 10 (3)
      13
    - 12 |  4 (3)
       1 | 24 (3)
```

2 Copy and complete these:
   a 4) 95   b 2) 57   c 3) 68

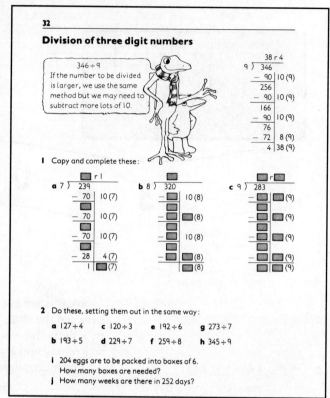

32

Division of three digit numbers

346 ÷ 9
If the number to be divided is larger, we use the same method but we may need to subtract more lots of 10.

```
          38 r 4
  9 ) 346
    - 90  | 10 (9)
     256
    - 90  | 10 (9)
     166
    - 90  | 10 (9)
      76
    - 72  |  8 (9)
       4  | 38 (9)
```

1 Copy and complete these:
   a 7) 239   b 8) 320   c 9) 283

2 Do these, setting them out in the same way:
   a 127 ÷ 4   c 120 ÷ 3   e 192 ÷ 6   g 273 ÷ 7
   b 193 ÷ 5   d 229 ÷ 7   f 259 ÷ 8   h 345 ÷ 9

   i 204 eggs are to be packed into boxes of 6.
      How many boxes are needed?
   j How many weeks are there in 252 days?

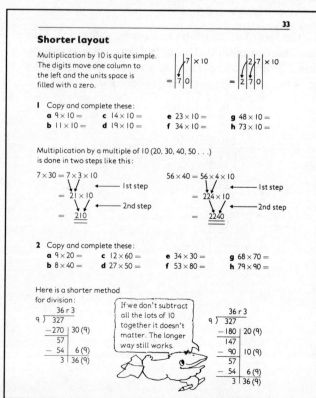

33

Shorter layout

Multiplication by 10 is quite simple. The digits move one column to the left and the units space is filled with a zero.

1 Copy and complete these:
   a 9 × 10 =   c 14 × 10 =   e 23 × 10 =   g 48 × 10 =
   b 11 × 10 =   d 19 × 10 =   f 34 × 10 =   h 73 × 10 =

Multiplication by a multiple of 10 (20, 30, 40, 50 . . .) is done in two steps like this:

7 × 30 = 7 × 3 × 10 ← 1st step
       = 21 × 10   ← 2nd step
       = 210

56 × 40 = 56 × 4 × 10 ← 1st step
        = 224 × 10   ← 2nd step
        = 2240

2 Copy and complete these:
   a 9 × 20 =   c 12 × 60 =   e 34 × 30 =   g 68 × 70 =
   b 8 × 40 =   d 27 × 50 =   f 53 × 80 =   h 79 × 90 =

Here is a shorter method for division:

```
        36 r 3
  9 ) 327
    - 270 | 30 (9)
       57
     - 54 |  6 (9)
        3 | 36 (9)
```

If we don't subtract all the lots of 10 together it doesn't matter. The longer way still works.

```
        36 r 3
  9 ) 327
    - 180 | 20 (9)
      147
     - 90 | 10 (9)
       57
     - 54 |  6 (9)
        3 | 36 (9)
```

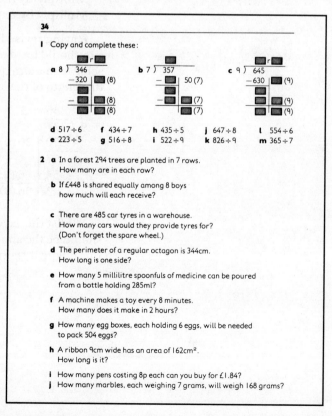

34

1 Copy and complete these:
   a 8) 346   b 7) 357   c 9) 645

   d 517 ÷ 6   f 434 ÷ 7   h 435 ÷ 5   j 647 ÷ 8   l 554 ÷ 6
   e 223 ÷ 5   g 516 ÷ 8   i 522 ÷ 9   k 826 ÷ 9   m 365 ÷ 7

2 a In a forest 294 trees are planted in 7 rows.
      How many are in each row?

   b If £448 is shared equally among 8 boys
      how much will each receive?

   c There are 485 car tyres in a warehouse.
      How many cars would they provide tyres for?
      (Don't forget the spare wheel.)

   d The perimeter of a regular octagon is 344cm.
      How long is one side?

   e How many 5 millilitre spoonfuls of medicine can be poured
      from a bottle holding 285ml?

   f A machine makes a toy every 8 minutes.
      How many does it make in 2 hours?

   g How many egg boxes, each holding 6 eggs, will be needed
      to pack 504 eggs?

   h A ribbon 9cm wide has an area of 162cm².
      How long is it?

   i How many pens costing 8p each can you buy for £1.84?

   j How many marbles, each weighing 7 grams, will weigh 168 grams?

Practice 7
Chapter 7: Division 1

Name _____

A

Complete these:

```
    97  r  5          81  r  1          78  r  0
6) 587             8) 649             7) 546
  -540  90 (6)       -640  80 (8)       -490  70 (7)
    47                  9                  56
  - 42   7 (6)        -  8    (8)        - 56    8 (7)
     5  97 (6)           81 (8)             0    78 (7)
```

B

Do these in the same way:

148÷4	37	736÷3	245 r 1	536÷6	89 r 2
736÷8	92	429÷5	85 r 4	847÷9	94 r 1

C

How many boxes, each holding 6 cakes,
will be needed to pack 468 cakes?   78

A prism has a volume of 333 cm³.
If its length is 9 cm what is the area of its cross-section?   37   cm²

A bag of nuts weighs 272 grams.
If each nut weighs 8 gms how many nuts are there?   34

If 8 girls share £7·36 equally between them each girl receives   £0·92

If the perimeter of a regular hexagon is 414 cm
how long is each side?   69 cm

How many pencils costing 7p each can I buy for £5·39?   77

Nuffield Maths 5 Spiritmasters   *Nuffield Maths 5 Pupils' Book, pages 31–34*   **20**

References and resources

Nuffield Mathematics Teaching Project, *Computation and Structure* ⑤
  Nuffield Guide, Chambers/Murray 1967 (See Introduction, page xi.)

Williams, E. M. and Shuard, H. *Primary Mathematics Today* Third
  Edition (Chapter 16), Longman Group Ltd 1982

Length 1

For the teacher

The aim of this chapter is to revise briefly some of the work covered in *Nuffield Maths 4 Pupils' Book* and to introduce multiplication and division of metric measures of length by a single digit.

The system of metric measures is so constructed that any physical quantity may be expressed in terms of *one unit only* with a decimal point separating whole units from parts of a unit. *Pupils' Book 4* introduced the idea of recording length in terms of metres and parts of the metre on a decimal abacus. Many children, however, will find difficulty in grasping the idea behind the decimal structure unless this gradual approach through metres, decimetres and centimetres is adhered to in the early stages.

The important skill of estimation is also emphasised so that children are encouraged to think about what sort of size they can reasonably expect an answer to be *before* actually carrying out computation.

Summary of the stages

1 Revision of metres and parts of a metre
2 Multiplication of metres and centimetres by a single digit
3 Division of metres and centimetres by a single digit

Vocabulary

Centimetre (cm), decimetre (dm), metre (m), parts of, tenth, hundredth, estimate, perimeter square, regular pentagon, hexagon, octagon, equilateral triangle, approximately equal to ($\simeq$), symmetrical.

Equipment and apparatus

Metre rules calibrated in centimetres, large measuring tape with metre, decimetre and centimetre markings, column abaci.

Working with the children

1 Revision of metres and parts of a metre

So far the relationships between a whole metre and parts of a metre have been emphasised by using common fractions, for example $1 \text{ dm} = \frac{1}{10} \text{ m}$ and $1 \text{ cm} = \frac{1}{100} \text{ m}$. This is revised briefly and then extended to decimal fraction notation:

$$1 \text{ dm} = \frac{1}{10} \text{ m} = 0.10 \text{ m} \qquad 1 \text{ cm} = \frac{1}{100} \text{ m} = 0.01 \text{ m}$$

At this point teachers may decide on a gradual change of headings for the abacus columns from

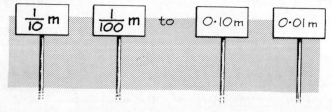

The children have now taken the step-by-step transition in the way of looking at parts of a metre a stage further:

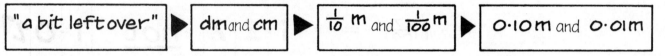

"a bit left over" ▶ dm and cm ▶ $\frac{1}{10}$ m and $\frac{1}{100}$ m ▶ 0·10 m and 0·01 m

The use of the decimetre as a *temporary teaching unit* serves as a link between metres and centimetres, emphasises the decimal structure of the system and facilitates the recording of lengths involving metres and centimetres to support the concept of place value.

A measuring tape, marked in m, dm and cm can be used as further reinforcement:

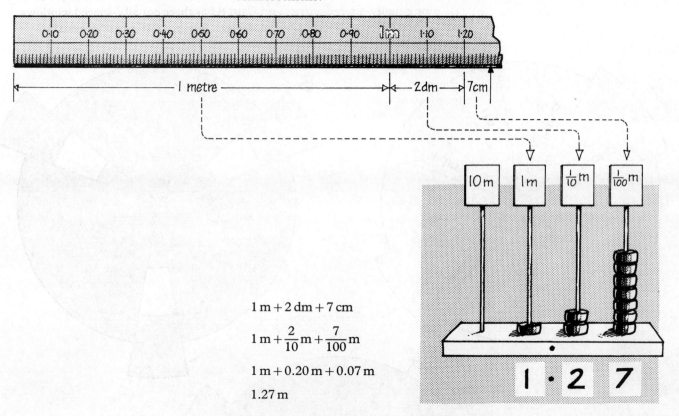

$1\,\text{m} + 2\,\text{dm} + 7\,\text{cm}$

$1\,\text{m} + \frac{2}{10}\,\text{m} + \frac{7}{100}\,\text{m}$

$1\,\text{m} + 0.20\,\text{m} + 0.07\,\text{m}$

$1.27\,\text{m}$

Some children may need to be reminded of earlier work on equivalent fractions, for example:

$$\frac{2}{10} = \frac{20}{100} = 0.20$$

This will help them to appreciate other relationships such as:

$\frac{2}{10} + \frac{7}{100}$ → $\frac{20}{100} + \frac{7}{100} = \frac{27}{100}$ → 0·27

0·20 + 0·07

45

and, at a later stage, will reduce any tendency to 'squeeze the 27 into the hundredths column'.

$$1\tfrac{27}{100} = \boxed{1 \cdot 2 \, 7} \ \underline{\textbf{not}} \ \boxed{1 \cdot 0 \, 27}$$

Note that 1.27 should be read as 'one point two seven', **not** as 'One point twenty-seven'.

A simple device made from two pieces of different coloured card helps to link the alternative versions of the same measure. Use a circular protractor as a template to mark 20° intervals on both cards. Card A has nine equally-spaced entries and Card B has three equally-spaced windows to show associated facts simultaneously. A paper fastener or press-stud joins the two cards at their centres.

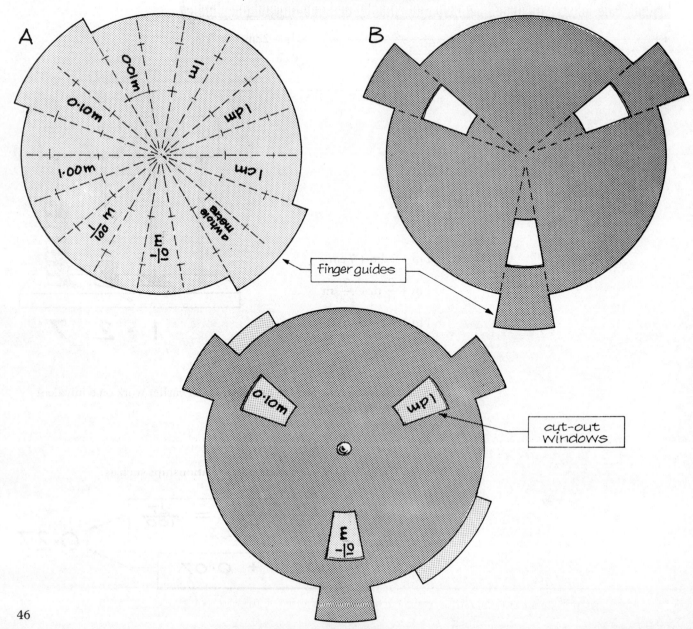

finger guides

cut-out windows

2 Multiplication of metres and centimetres by a single digit

In *Nuffield Maths 4* multiplication by 2 was introduced as a shorter method for finding perimeters of rectangles, and multiplication by 4 hinted at for the perimeters of squares. The development of this multiplication process will consider the calculation of perimeters of equilateral triangles, regular pentagons, hexagons and octagons.

The point was made in *Nuffield Maths 3 Teachers' Handbook* that, unfortunately, most commercially-produced sets of polygons are *regular*, that is with equal sides and equal angles. This often gives the impression that the name pentagon, for example, only applies to regular pentagons. The first examples involve finding the perimeters of regular polygons by multiplying the length of one side by the number of sides. Later examples use polygons which, although they have equal sides, are not *regular* polygons because their angles are not equal. It is not intended that work on the multiplication of metric measurement should be restricted to finding the perimeters of plane figures. Other problems such as finding the total height of a stack of cartons or the length of a row of chairs should be presented to children in the hope that they will develop the ability to think up problems for themselves.

Again, the importance of estimation before calculation is stressed. 'Think before ink' is not as silly as it sounds. If a child can look at a regular hexagon with sides of 19 cm and think of 'six lots of 20' before attempting to calculate the perimeter, he is well on the way to success. Estimating procedure may vary from child to child but it is important to encourage the ability to 'think round the problem' first and to record in some way the estimated answer. Of course the whole point of estimation is lost if there is no comparison made between the estimate and the calculation result – 'Was my estimate close? If not, have I gone wrong in the calculation?'

When dealing with dimensions given in metres and decimal parts of a metre, it is essential that decimal points are kept in line, one beneath the other. This is emphasised in the first instance by using an example involving the addition of the lengths of sides of a scalene quadrilateral.

For multiplying decimal quantities by a single digit, 'keeping the points in line' is again emphasised. Which of the two layouts is used – the 'partial products' or the shorter 'telescoped' format – will depend on the level of ability and confidence of the child.

Partial products		*Shorter layout*
2.73		2.73
× 6		× 6
.18	(.03 × 6)	16.38
4.20	(.70 × 6)	4 1
12.00	(2.00 × 6)	
16.38		

3 Division of m and cm by a single digit

Of the four operations, that of division is probably the least-needed, particularly in the case of measurement of length. It is also a very difficult operation for some children. At all times, whatever the operation, there is a real need for children to be presented with everyday problems involving measurement which they can understand. Often, the real difficulty lies in an inability to decide which operation to use in the first place.

A first step to division by a single digit could be the checking of previous work on multiplication by a single digit. Children should develop an awareness of the possible uses of inverse operations as checks.

For example, an answer arrived at by subtraction may be checked by the correct application of addition. The same goes for the operations of multiplication and division.

In *Pupils' Book 4* the children were asked to find a quick way of working out the perimeter of a square. Most children will eventually decide to multiply 32 cm by 4, to give the perimeter as 128 cm or 1.28 m.

Now the problem is: If a square has a perimeter of 1.28 m find the length of each side in cm.

Similar problems are given for other regular polygons including the heptagon (7 sides) and nonagon (9 sides).

By changing the length to be divided to cm first, the calculation is done without decimal points. For example:

$13.80 \text{ m} \div 4 = 1380 \text{ cm} \div 4$

```
        345 cm
4 ) 1380 cm
  − 1200        300 (4)
     180
  −  160         40 (4)
      20
  −   20          5 (4)
                345
```

This method is not applicable to all division of decimal quantities by a single whole number but, as with the division of £ and p, it acts as a 'gentle introduction'.

Alternatively, the inverse of the 'partial products' method of multiplication can be used, provided the decimal points are kept in line:

```
         3.45 m
4 ) 13.80 m
    12.00    3.00 × 4
     1.80
     1.60    0.4 × 4
      .20
      .20    0.05 × 4
            3.45
```

When using this approach it has been found advisable to replace the bracket notation, 0.05(4), by 0.05 × 4. This avoids the awkward phrase 'Nought-point-nought-five lots of four' and it is more realistic to think of 0.05 of a metre being multiplied by 4 than *vice versa*.

Examples using unit fractions remind children that 'Find $\frac{1}{5}$ of . . . .' is an alternative form of words for 'Divide by 5'.

The importance of estimation as a check on the reasonableness of a calculated result is emphasised yet again. Every effort should be made to ensure that children habitually follow the dictum:

First estimate, then calculate, then compare.

Pages from the Pupils' Book and Spiritmasters

Chapter 8: Length 1

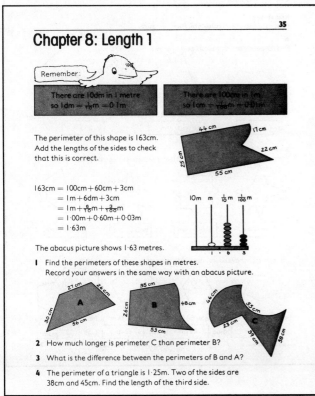

Remember:

There are 10dm in 1 metre
so $1dm = \frac{1}{10}m = 0.1m$

There are 100cm in 1m
so $1cm = \frac{1}{100}m = 0.01m$

The perimeter of this shape is 163cm.
Add the lengths of the sides to check
that this is correct.

$$163cm = 100cm + 60cm + 3cm$$
$$= 1m + 6dm + 3cm$$
$$= 1m + \frac{6}{10}m + \frac{3}{100}m$$
$$= 1.00m + 0.60m + 0.03m$$
$$= 1.63m$$

The abacus picture shows 1.63 metres.

1 Find the perimeters of these shapes in metres.
   Record your answers in the same way with an abacus picture.

2 How much longer is perimeter C than perimeter B?

3 What is the difference between the perimeters of B and A?

4 The perimeter of a triangle is 1.25m. Two of the sides are
   38cm and 45cm. Find the length of the third side.

Multiplication of metres and centimetres

An **equilateral** triangle has all its sides the same length.
To find the perimeter we can multiply the length of one side by 3.

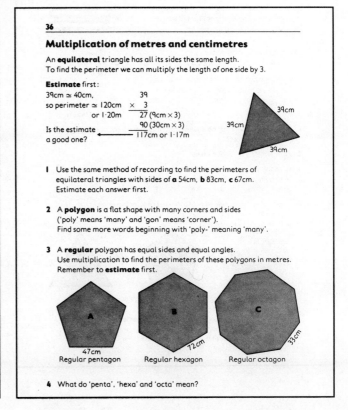

Estimate first:
$39cm \simeq 40cm$
so perimeter $\simeq 120cm$
or $1.20m$

$$\begin{array}{r} 39 \\ \times \quad 3 \\ \hline 27 \ (9cm \times 3) \\ 90 \ (30cm \times 3) \\ \hline 117cm \ or \ 1.17m \end{array}$$

Is the estimate
a good one? ← 117cm or 1.17m

1 Use the same method of recording to find the perimeters of
   equilateral triangles with sides of **a** 54cm, **b** 83cm, **c** 67cm.
   Estimate each answer first.

2 A **polygon** is a flat shape with many corners and sides
   ('poly' means 'many' and 'gon' means 'corner').
   Find some more words beginning with 'poly-' meaning 'many'.

3 A **regular** polygon has equal sides and equal angles.
   Use multiplication to find the perimeters of these polygons in metres.
   Remember to **estimate** first.

47cm
Regular pentagon

72cm
Regular hexagon

33cm
Regular octagon

4 What do 'penta', 'hexa' and 'octa' mean?

1 These polygons have equal sides but they are not regular.
   a Why not?   **b** Are they symmetrical?
   c First estimate and then work out their perimeters in metres.

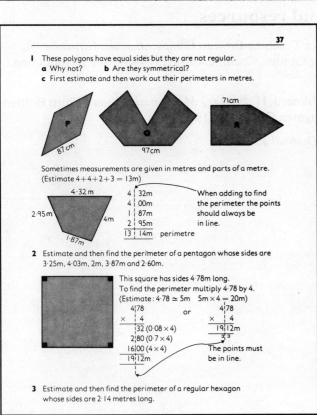

87cm   97cm   71cm

Sometimes measurements are given in metres and parts of a metre.
(Estimate $4 + 4 + 2 + 3 = 13m$)

4.32 m
2.95 m
4m
1.87m

$$\begin{array}{r} 4\ \ 32m \\ 4\ \ 00m \\ 1\ \ 87m \\ 2\ \ 95m \\ \hline 13\ \ 14m \quad perimetre \end{array}$$

When adding to find
the perimeter the points
should always be
in line.

2 Estimate and then find the perimeter of a pentagon whose sides are
   3.25m, 4.03m, 2m, 3.87m and 2.60m.

This square has sides 4.78m long.
To find the perimeter multiply 4.78 by 4.
(Estimate: $4.78 \simeq 5m$ $5m \times 4 = 20m$)

$$\begin{array}{r} 4\ \ 78 \\ \times \quad 4 \\ \hline \ \ 32 \ (0.08 \times 4) \\ 2\ \ 80 \ (0.7 \times 4) \\ 16\ \ 00 \ (4 \times 4) \\ \hline 19\ \ 12m \end{array}$$

or

$$\begin{array}{r} 4\ \ 78 \\ \times \quad 4 \\ \hline 19\ \ 12m \\ 3\ \ 3 \end{array}$$

The points must
be in line.

3 Estimate and then find the perimeter of a regular hexagon
   whose sides are 2.14 metres long.

Division of metres and centimetres

The perimeter of this square is 1.28m.
The sides are equal in length so to find
the length of one side, we divide the perimeter by 4.

$1.28m \div 4 = 128cm \div 4$

(Estimate: 'just over $120' \div 4$
is 'just over 30',
so length of side $\simeq$ 30cm.)

Length of side = 32cm or 0.32m.

$$\begin{array}{r} 32 \\ 4\ )\ \overline{128} \\ -120 \quad 30\ (4) \\ \hline 8 \\ -\ 8 \quad 2\ (4) \\ \hline 32 \end{array}$$

1 Find the lengths of sides of squares with these perimeters.
   Give each answer first in cm then in m.
   a 1.24m   **b** 2.88m   **c** 7.64m   **d** 4.04m   **e** 5.68m

2 For each of these regular polygons find the length of one side.
   Give each answer first in cm then in m.

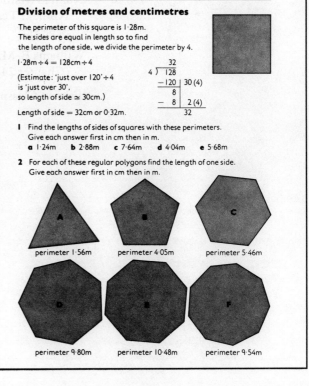

perimeter 1.56m   perimeter 4.05m   perimeter 5.46m

perimeter 9.80m   perimeter 10.48m   perimeter 9.54m

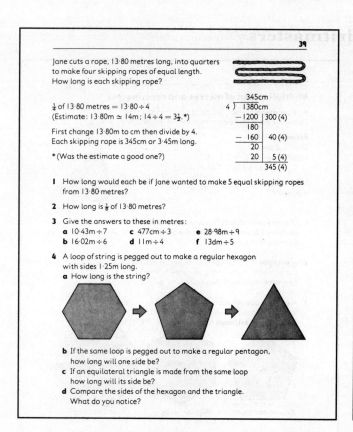

39

Jane cuts a rope, 13·80 metres long, into quarters to make four skipping ropes of equal length. How long is each skipping rope?

$\frac{1}{4}$ of 13·80 metres = 13·80 ÷ 4
(Estimate: 13·80m ≃ 14m; 14 ÷ 4 = 3½.*)

First change 13·80m to cm then divide by 4.
Each skipping rope is 345cm or 3·45m long.

* (Was the estimate a good one?)

```
                    345cm
          4 ) 1380cm
            − 1200 | 300 (4)
               180
            −  160 |  40 (4)
                20
                20 |   5 (4)
               ──────────────
                    345 (4)
```

1 How long would each be if Jane wanted to make 5 equal skipping ropes from 13·80 metres?

2 How long is $\frac{1}{8}$ of 13·80 metres?

3 Give the answers to these in metres :
  a 10·43m ÷ 7   **c** 477cm ÷ 3   **e** 28·98m ÷ 9
  b 16·02m ÷ 6   **d** 11m ÷ 4   **f** 13dm ÷ 5

4 A loop of string is pegged out to make a regular hexagon with sides 1·25m long.
  a How long is the string?

  b If the same loop is pegged out to make a regular pentagon, how long will one side be?
  c If an equilateral triangle is made from the same loop how long will its side be?
  d Compare the sides of the hexagon and the triangle. What do you notice?

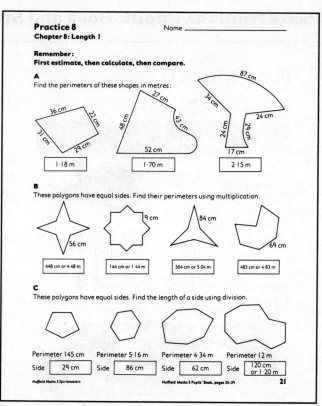

Practice 8
Chapter 8: Length 1

Name _____

Remember:
First estimate, then calculate, then compare.

A
Find the perimeters of these shapes in metres:

36 cm 31 cm 22 cm 29 cm | 1·18 m |

27 cm 48 cm 43 cm 52 cm | 1·70 m |

87 cm 34 cm 24 cm 24 cm 29 cm 17 cm | 2·15 m |

B
These polygons have equal sides. Find their perimeters using multiplication.

56 cm | 448 cm or 4·48 m |

9 cm | 144 cm or 1·44 m |

84 cm | 504 cm or 5·04 m |

69 cm | 483 cm or 4·83 m |

C
These polygons have equal sides. Find the length of a side using division.

Perimeter 145 cm   Side | 29 cm |

Perimeter 5·16 m   Side | 86 cm |

Perimeter 4·34 m   Side | 62 cm |

Perimeter 12 m   Side | 120 cm or 1·20 m |

Nuffield Maths 5 Spiritmasters   *Nuffield Maths 5 Pupils' Book, pages 35–39*   **21**

References and resources

Nuffield Mathematics Teaching Project, *Computation and Structure* ⑤, Nuffield Teaching Guides, Chambers/Murray 1967 (See Introduction, page xi.)

Williams E. M. and Shuard, H. *Primary Mathematics Today* Third Edition (Chapter 24), Longman Group Ltd 1982

Triman Classmate, *Decimal abacus*

For the teacher

In this chapter the pupil's experience of calculating, using money, is extended to multiplication and division. The opportunity is taken to reinforce multiplication and division of decimals by a single digit. The techniques used in this chapter will be met again both with other measurements and with pure number.

Summary of the stages

1 Revision of previous work
2 Multiplication of £ by a single digit
3 Division of £ by a single digit
4 Money in shopping problems

Vocabulary

Amount, beneath, decimal point, column.

Equipment and apparatus

Catalogues, advertisements.

Working with the children

1 Revision of previous work

The first four questions in this chapter revise the work previously covered – the writing of amounts given in pence as pounds and vice versa, and addition and subtraction of amounts less than £10.

The main emphases in these questions are placed on the importance of the position of the decimal point and the setting down of amounts with the decimal points beneath each other – both of paramount importance in the work to follow.

Note: amounts of money should be recorded using *either* '£' or 'p' but *not both*. For example: £3.45 or 345p but *not* £3.45p.

2 Multiplication of £ by a single digit

The method of multiplication used in *Pupils' Book 5* is:

Set down the figures and place the decimal point in answer line below that in the amount to be multiplied.

$$\begin{array}{r} £3.62 \\ \times\ \ 7 \\ \hline . \\ \end{array}$$

The multiplication is then carried out as normal and the point is in the correct place:

$$\begin{array}{r} £3.62 \\ \times\ \ 7 \\ \hline £25.34 \\ \end{array}$$

It is not breaking our place value rules to put the 7 beneath the 2 hundredths. The £3.62 is an *amount of money* and the 7 represents the *number of items* for which we are finding the total cost.

3 Division of £ by a single digit

The method used here for division will be developed later as the one for all division of decimals. In view of this its importance cannot be stressed too greatly. As division is the inverse of multiplication, this method may be seen as the inverse of the method used for multiplication. There the decimal point was fixed *before* any working was done, here it is fixed *after* all the working is done.

£8.96 ÷ 7 is set down and worked out *without* the decimal point.

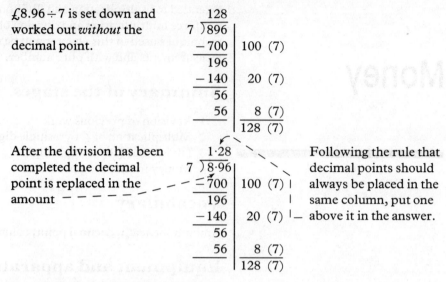

After the division has been completed the decimal point is replaced in the amount – – – – –

Following the rule that decimal points should always be placed in the same column, put one above it in the answer.

4 Money in shopping problems

These pages give practice covering all work previously done. Some are reasonably simple but others involve three or four calculations. (*Nuffield Maths 5 Spiritmasters*, Grid 11.)

For further practice, or to extend more able children, teachers may wish to make their own sets of questions based on pages from catalogues.

Pages from the Pupils' Book and Spiritmasters

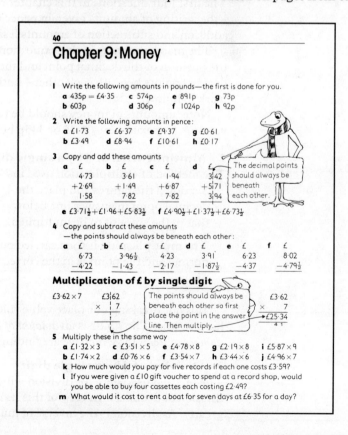

40

Chapter 9: Money

1 Write the following amounts in pounds—the first is done for you.
  a 435p = £4·35   **c** 574p   **e** 891p   **g** 73p
  b 603p   **d** 306p   **f** 1024p **h** 92p

2 Write the following amounts in pence:
  a £1·73   **c** £6·37   **e** £9·37   **g** £0·61
  b £3·49   **d** £8·94   **f** £10·61 **h** £0·17

3 Copy and add these amounts

a £	**b** £	**c** £	**d** £
4·73	3·61	1·94	3·42
+2·69	+1·49	+6·87	+5·71
1·58	7·82	7·82	3·94

The decimal points should always be beneath each other.

  e £3·71½+£1·96+£5·83½   **f** £4·90½+£1·37½+£6·73½

4 Copy and subtract these amounts
  —the points should always be beneath each other:

a £	**b** £	**c** £	**d** £	**e** £	**f** £
6·73	3·96½	4·23	3·91	6·23	8·02
−4·22	−1·43	−2·17	−1·87½	−4·37	−4·79½

Multiplication of £ by single digit

£3·62 × 7

The points should always be beneath each other so first place the point in the answer line. Then multiply.

£3·62 × 7 → £25·34

5 Multiply these in the same way
  a £1·32 × 3   **c** £3·51 × 5   **e** £4·78 × 8   **g** £2·19 × 8   **i** £5·87 × 9
  b £1·74 × 2   **d** £0·76 × 6   **f** £3·54 × 7   **h** £3·44 × 6   **j** £4·96 × 7
  k How much would you pay for five records if each one costs £3·59?
  l If you were given a £10 gift voucher to spend at a record shop, would you be able to buy four cassettes each costing £2·49?
  m What would it cost to rent a boat for seven days at £6·35 for a day?

52

Division by single digits

£8·96 ÷ 7

We write the amount without its decimal point and divide as we have done before:

```
        128
    7 ) 896
      - 700   100 (7)
        196
      - 140    20 (7)
         56
      -  56     8 (7)
              128 (7)
```

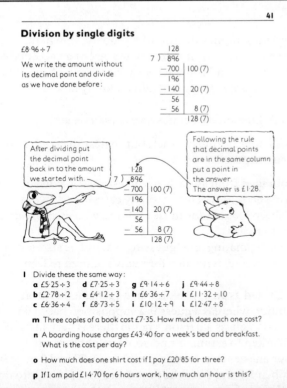

After dividing put the decimal point back in to the amount we started with.

Following the rule that decimal points are in the same column put a point in the answer. The answer is £1·28.

```
         1·28
    7 ) 896
      - 700   100 (7)
        196
      - 140    20 (7)
         56
      -  56     8 (7)
              128 (7)
```

1 Divide these the same way:
- **a** £5·25 ÷ 3
- **b** £2·78 ÷ 2
- **c** £6·36 ÷ 4
- **d** £7·25 ÷ 3
- **e** £4·12 ÷ 3
- **f** £8·73 ÷ 5
- **g** £9·14 ÷ 6
- **h** £6·36 ÷ 7
- **i** £10·12 ÷ 9
- **j** £9·44 ÷ 8
- **k** £11·32 ÷ 10
- **l** £12·47 ÷ 8

m Three copies of a book cost £7·35. How much does each one cost?

n A boarding house charges £43·40 for a week's bed and breakfast. What is the cost per day?

o How much does one shirt cost if I pay £20·85 for three?

p If I am paid £14·70 for 6 hours work, how much an hour is this?

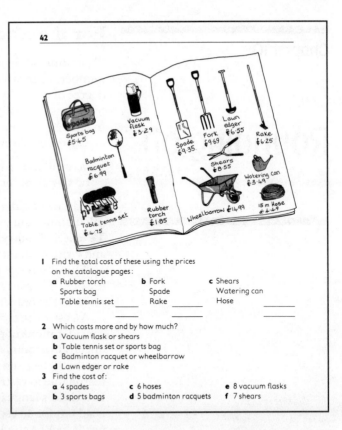

1 Find the total cost of these using the prices on the catalogue pages:
- **a** Rubber torch / Sports bag / Table tennis set _____
- **b** Fork / Spade / Rake _____
- **c** Shears / Watering can / Hose _____

2 Which costs more and by how much?
- **a** Vacuum flask or shears
- **b** Table tennis set or sports bag
- **c** Badminton racquet or wheelbarrow
- **d** Lawn edger or rake

3 Find the cost of:
- **a** 4 spades
- **b** 3 sports bags
- **c** 6 hoses
- **d** 5 badminton racquets
- **e** 8 vacuum flasks
- **f** 7 shears

1 Copy and complete the following bills:

- **a** 4 spades / 3 forks / 5 lawn edgers _____ / Total _____
- **b** 5 table tennis sets / 7 badminton racquets / 4 sports bags / Total _____
- **c** 8 torches / 6 hoses / 5 watering cans _____ / Total _____
- **d** 3 wheelbarrows / 6 shears / 5 rakes / Total _____

2 If I buy 3 torches and a vacuum flask, how much change should I get from a £10 note?

3 Here is a bill from last year. What was the price last year for:
- **a** one torch.
- **b** one vacuum flask.
- **c** a sports bag.
- **d** one badminton racquet?

4 How much dearer is each item in the bill this year?

5 Make out a similar bill using this year's prices. How much more is the total this year?

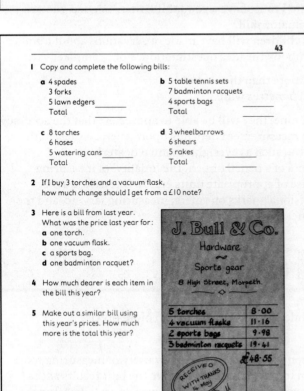

J. Bull & Co.
Hardware
~
Sports gear
8 High Street, Morpeth.

5 torches	8·00
4 vacuum flasks	11·16
2 sports bags	9·98
3 badminton racquets	19·41
	£48·55

RECEIVED WITH THANKS 14 May
J Bull

Practice 9
Chapter 9: Money

Name _____

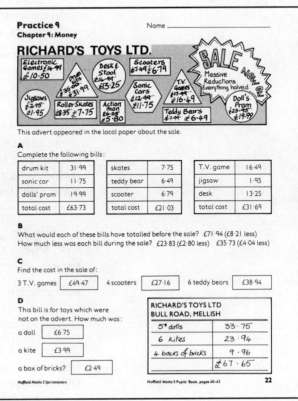

RICHARD'S TOYS LTD.

This advert appeared in the local paper about the sale.

A

Complete the following bills:

drum kit	31·99
sonic car	11·75
dolls' pram	19·99
total cost	£63·73

skates	7·75
teddy bear	6·49
scooter	6·79
total cost	£21·03

T.V. game	16·49
jigsaw	1·95
desk	13·25
total cost	£31·69

B

What would each of these bills have totalled before the sale? £71·94 (£8·21 less)
How much less was each bill during the sale? £23·83 (£2·80 less)   £35·73 (£4·04 less)

C

Find the cost in the sale of:

3 T.V. games [£49·47]   4 scooters [£27·16]   6 teddy bears [£38·94]

D

This bill is for toys which were not on the advert. How much was:

a doll [£6·75]

a kite [£3·99]

a box of bricks? [£2·49]

RICHARD'S TOYS LTD BULL ROAD, MELLISH		
5 dolls	33·75	
6 kites	23·94	
4 boxes of bricks	9·96	
	£67·65	

Chapter 10

Rounding off

For the teacher

'Rounding off' in mathematics implies the rounding 'up' or 'down' of numbers or measures in order to arrive at values which are of an acceptable degree of accuracy for discussion and comparison at a particular time. The process is generally known as 'approximation' and is closely linked with that of 'estimation'.

Chambers 20th Century Dictionary defines *approximation* as:

> 'a result in mathematics not rigorously exact, but so near the truth as to be sufficient for a given purpose.'

Children should be encouraged from an early age to estimate results of calculations and measuring activities before actually carrying them out. Teachers should keep a sharp lookout for those children who measure first and write in an estimate afterwards!

The difference between estimating and 'guessing' should be stressed at the outset because there is a strong tendency for many children to lean towards the latter.

An estimate can be regarded as an approximate judgement based on considerations of probability, whereas a guess is a supposition based on uncertain grounds. If asked to guess a person's name there is very little, if anything, to go on; but if asked to estimate a person's height it would be reasonable to restrict your answer to something between 1.20 and 2 metres. In a child's language, 'When I guess I say the first thing that comes into my head but for an estimate I must have a good think first.' A good guesser needs luck; a good estimator skill.

In their early years children will have made observations about number and quantity which embody the idea of estimation. For example:

'The cup will hold more than the beaker.'
'This room is about 5 metres wide.'

Perhaps, by the age of nine, they will be able to appreciate that the accuracy (or near accuracy) of a measurement depends upon several issues, both physical and mechanical, such as eyesight, manual dexterity, co-ordination, thickness of a pencil or pen, the 'quality' of measuring instruments and the use of appropriate units etc. In this case 'quality' includes clarity of divisional marks on rulers, measuring jars etc. and the correct placing of numerals:

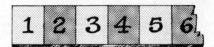

misleading to many children

much more precise

It is only through continued practice with a variety of measuring tools that children will learn to select those which give the best results at the time. There is much scope for teacher/class discussion on 'Choosing the right tool for the job' for all aspects of measurement.

While all this is going on, the children should learn that however suitable the tool used, there will always be an element of approximation about the result. Measurement can never be absolutely accurate.

Summary of the stages

1 The number line and similar graduated scales
2 Rounding off decimal numbers
3 Approximate area
4 Approximate volume

Vocabulary

Approximate, approximation, estimate, equally spaced, nearest whole number, centimetre, metre, hour, etc., half-way, rounding off/up/down, graduated.

Equipment and apparatus

Number line, squared paper, graduated measuring jars, clock faces (rubber stamps), rulers marked in half centimetres or in millimetres, measuring tapes.

Working with the children

1   The number line and similar graduated scales

Some revision of earlier work on the number line will be beneficial especially if it emphasises that the whole-number markings are *equally spaced*. The importance of the half-way mark as the deciding line between rounding up and rounding down needs to be introduced carefully. It is not always obvious to some children that a point to the left (that is below) the half-way mark is nearer to the lower than to the higher whole number.

The rule: under half-way, round **down**
half-way and over, round **up**

must be established and adhered to throughout. This may prompt the comment from some children, 'Why is half-way always rounded up? It's not fair because half-way is just as near to the lower number as it is to the higher number.' It should be explained that readings on the half-way mark have to go one way or the other and it is much more sensible if everyone obeys the same rule. In any case, it is really quite 'fair' because the size of the section of number line rounded off to each number is the same.

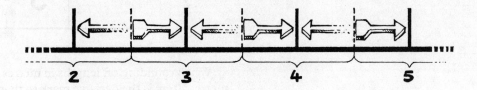

Several versions of the 'approximately equal to' sign are in use: $\approx \eqsim \simeq$. Most children are able to draw the last fairly easily so this has been adopted in the Pupils' Book. The important thing is that the appearance of the symbol 'nearly an = sign' should be a reminder of its meaning.

The experience gained on the number line can be used for reading other equally-spaced graduations. Even if rulers marked in half-centimetres or 5-millimetre intervals are not available, most children will be able to make a visual judgement on whether to round up or down by imagining the half-way mark. (*Nuffield Maths 5 Spiritmasters*, Grid 29.)

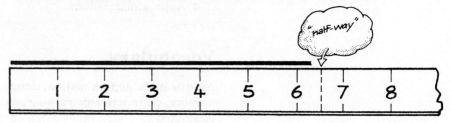

In the absence of graduated cylinders, the idea suggested in *Nuffield Maths 4 Teachers' Book* can be adapted to give extra practice in reading vertical scales. A cardboard scale is pinned to the edge of a blackboard and children take turns in chalking in and reading off 'columns of water' to the nearest 100 ml.

One of the most common uses of approximation is probably in the reading of clock faces. Phrases such as 'nearly ten past' are often used by children and the exercises use the familiar clock face to introduce the idea of approximating on a circular scale. Other examples are parking meters, bathroom scales, weighing machines, etc.

2 Rounding off decimal numbers

The idea of using the half-way mark to decide whether to round up or down is still used, it is only the notation which is changed. Half-way is now $\frac{5}{10}$ or 0.5 when approximating to the nearest whole number. This has already been met by children using rulers marked in cm and mm where the 5 mm mark, that is .5 cm, was the deciding line. A start is made by rounding off numbers with one decimal place but some of the later examples involve two or three decimal places. Children who are confused by this should be reminded that it is the '*tenths*' *digit* that is important when rounding off to the nearest whole number. The other digits in the hundredths, thousandths, etc. columns are ignored.

hundreds	tens	units	$\frac{1}{10}$ ths	$\frac{1}{100}$ ths	$\frac{1}{1000}$ th
		3	· 2	5	

When rounding off lengths in metres and centimetres to the nearest metre, 50 cm is the decisive mark so that 4 m 52 cm $\simeq$ 5 m and 3.49 m $\simeq$ 3 m. The exercise on the perimeter of the isosceles triangle illustrates the difference in results which arises from the timing of approximation. Rounding off the length of each side before adding gives a higher result in this case than adding first and then rounding off. Some children may be able to find cases where the opposite is true (for example 5.44, 5.44, 3.20) or where the approximate answers are the same (for example 5.20, 5.20, 3.70). Discussion on why these differences occur should lead to interesting insights on the children's thinking about approximation.

3 Approximate area

Rounding off the dimensions of rectangles in order to find an approximate area has obvious links with working out, at a later stage, approximate answers for multiplication of decimals (*Nuffield Maths 6 Pupils' Book*).

Children have already found approximate areas by counting squares. This method also rounded off by counting half a square or more as a whole square and ignoring less than half a square. Comparing the approximate results obtained by the two methods can lead to interesting discussion. For example, all three rectangles A B and C have the same area (10.08 cm²).

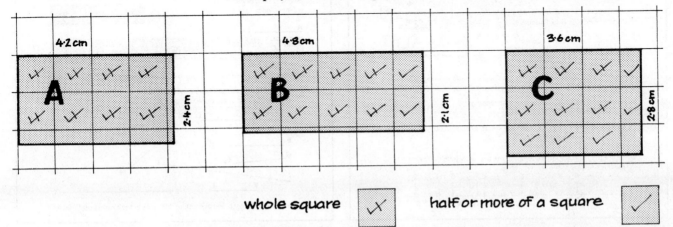

Approximate areas to the nearest cm², by counting squares, are:

A 8 cm²;   B 10 cm²;   C 11 cm².

Similar results are given by rounding off dimensions to the nearest cm and multiplying. The best approximation is B where one dimension was rounded up and the other down, $(4.8 \times 2.1) \simeq (5 \times 2)$.

The children should be encouraged to find further uses of approximation in area to supplement those given in the text – areas of lawns for seeding or treatment, areas covered by contents of tins of paint, etc. (*Nuffield Maths 5 Spiritmasters*, Grids 6 and 7.)

4 Approximate volume

This section follows the same format as approximate area. In the case of a cube the same dimension is rounded up or down three times so the difference between approximate and actual volume is likely to be greater. Some examples of cuboids involve both rounding up and rounding down of dimensions which will tend to have a compensatory effect on the approximation. If electronic calculators are available, some children will be able to calculate actual volumes for comparison with approximate results.

Once children are confident, other measures may be investigated. In the case of 'the nearest kilogram' or 'the nearest litre' the half-way marks are 500 grams and 500 millilitres respectively so that 5 kg and 380 g or 5.380 kg $\simeq$ 5 kg (to the nearest kg) and 2 litres and 700 ml or 2.700 litres $\simeq$ 3 litres (to the nearest litre).

Pages from the Pupils' Book and Spiritmasters

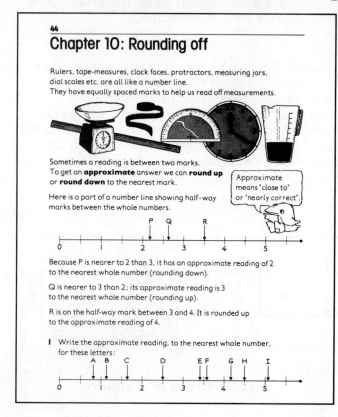

44

Chapter 10: Rounding off

Rulers, tape-measures, clock faces, protractors, measuring jars, dial scales etc. are all like a number line.
They have equally spaced marks to help us read off measurements.

Sometimes a reading is between two marks.
To get an **approximate** answer we can **round up** or **round down** to the nearest mark.

Here is a part of a number line showing half-way marks between the whole numbers.

Approximate means 'close to' or 'nearly correct'.

Because P is nearer to 2 than 3, it has an approximate reading of 2 to the nearest whole number (rounding down).

Q is nearer to 3 than 2; its approximate reading is 3 to the nearest whole number (rounding up).

R is on the half-way mark between 3 and 4. It is rounded up to the approximate reading of 4.

1 Write the approximate reading, to the nearest whole number, for these letters:

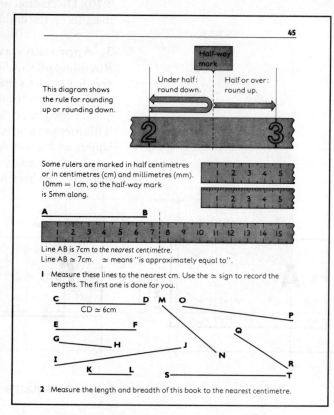

45

This diagram shows the rule for rounding up or rounding down.

Under half: round down. Half or over: round up.

Some rulers are marked in half centimetres or in centimetres (cm) and millimetres (mm). 10mm = 1cm, so the half-way mark is 5mm along.

Line AB is 7cm *to the nearest centimetre*.
Line AB ≃ 7cm.   ≃ means "is approximately equal to".

1 Measure these lines to the nearest cm. Use the ≃ sign to record the lengths. The first one is done for you.

CD ≃ 6cm

2 Measure the length and breadth of this book to the nearest centimetre.

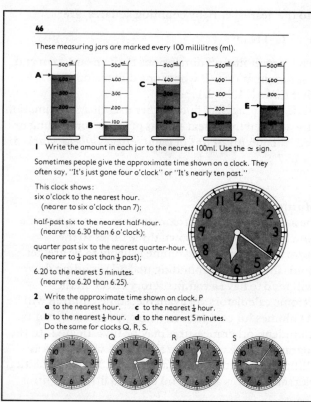

46

These measuring jars are marked every 100 millilitres (ml).

1 Write the amount in each jar to the nearest 100ml. Use the ≃ sign.

Sometimes people give the approximate time shown on a clock. They often say, "It's just gone four o'clock" or "It's nearly ten past."

This clock shows:
six o'clock to the nearest hour.
   (nearer to six o'clock than 7);

half-past six to the nearest half-hour.
   (nearer to 6.30 than 6 o'clock);

quarter past six to the nearest quarter-hour.
   (nearer to ¼ past than ½ past);

6.20 to the nearest 5 minutes.
   (nearer to 6.20 than 6.25).

2 Write the approximate time shown on clock, P
   a to the nearest hour.   **c** to the nearest ¼ hour.
   b to the nearest ½ hour.   **d** to the nearest 5 minutes.
   Do the same for clocks Q, R, S.

P   Q   R   S

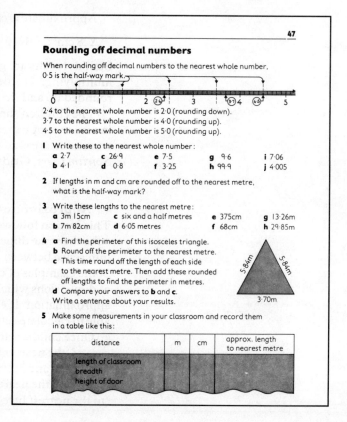

47

Rounding off decimal numbers

When rounding off decimal numbers to the nearest whole number, 0·5 is the half-way mark.

2·4 to the nearest whole number is 2·0 (rounding down).
3·7 to the nearest whole number is 4·0 (rounding up).
4·5 to the nearest whole number is 5·0 (rounding up).

1 Write these to the nearest whole number:
   a 2·7   **c** 26·9   **e** 7·5   **g** 9·6   **i** 7·06
   b 4·1   **d** 0·8   **f** 3·25   **h** 99·9   **j** 4·005

2 If lengths in m and cm are rounded off to the nearest metre, what is the half-way mark?

3 Write these lengths to the nearest metre:
   a 3m 15cm   **c** six and a half metres   **e** 375cm   **g** 13·26m
   b 7m 82cm   **d** 6·05 metres   **f** 68cm   **h** 29·85m

4 **a** Find the perimeter of this isosceles triangle.
   b Round off the perimeter to the nearest metre.
   c This time round off the length of each side to the nearest metre. Then add these rounded off lengths to find the perimeter in metres.
   d Compare your answers to **b** and **c**.
   Write a sentence about your results.

5·84m   5·84m
3·70m

5 Make some measurements in your classroom and record them in a table like this:

distance	m	cm	approx. length to nearest metre
length of classroom breadth height of door			

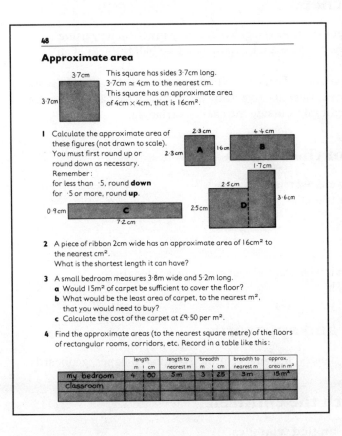

48

Approximate area

3·7cm

This square has sides 3·7cm long.
3·7cm ≃ 4cm to the nearest cm.
This square has an approximate area
of 4cm × 4cm, that is 16cm².

1 Calculate the approximate area of
these figures (not drawn to scale).
You must first round up or
round down as necessary.
Remember:
for less than ·5, round **down**
for ·5 or more, round **up**.

2 A piece of ribbon 2cm wide has an approximate area of 16cm² to
the nearest cm².
What is the shortest length it can have?

3 A small bedroom measures 3·8m wide and 5·2m long.
a Would 15m² of carpet be sufficient to cover the floor?
b What would be the least area of carpet, to the nearest m²,
that you would need to buy?
c Calculate the cost of the carpet at £9·50 per m².

4 Find the approximate areas (to the nearest square metre) of the floors
of rectangular rooms, corridors, etc. Record in a table like this:

	length m	cm	length to nearest m	·breadth m	cm	breadth to nearest m	approx. area in m²
my bedroom	4	80	5m	3	25	3m	15m²
classroom							

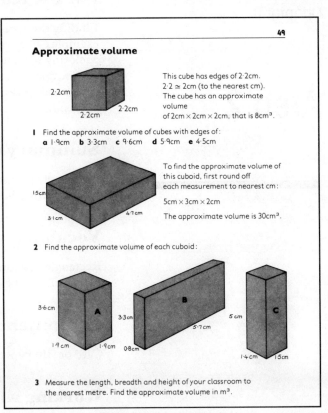

49

Approximate volume

2·2cm

This cube has edges of 2·2cm.
2·2 ≃ 2cm (to the nearest cm).
The cube has an approximate
volume
of 2cm × 2cm × 2cm, that is 8cm³.

1 Find the approximate volume of cubes with edges of:
a 1·9cm b 3·3cm c 9·6cm d 5·9cm e 4·5cm

To find the approximate volume of
this cuboid, first round off
each measurement to nearest cm:

5cm × 3cm × 2cm

The approximate volume is 30cm³.

2 Find the approximate volume of each cuboid:

3 Measure the length, breadth and height of your classroom to
the nearest metre. Find the approximate volume in m³.

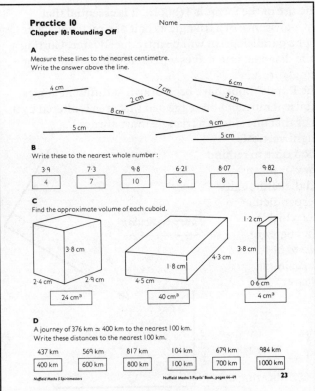

Practice 10
Chapter 10: Rounding Off   Name _____

A
Measure these lines to the nearest centimetre.
Write the answer above the line.

4 cm   7 cm   6 cm
2 cm   3 cm
8 cm
5 cm   9 cm
5 cm

B
Write these to the nearest whole number:

3·9	7·3	9·8	6·21	8·07	9·82
4	7	10	6	8	10

C
Find the approximate volume of each cuboid.

24 cm³	40 cm³	4 cm³

D
A journey of 376 km ≃ 400 km to the nearest 100 km.
Write these distances to the nearest 100 km.

437 km	569 km	817 km	104 km	679 km	984 km
400 km	600 km	800 km	100 km	700 km	1000 km

Nuffield Maths 5 Spiritmasters   *Nuffield Maths 5 Pupils' Book, pages 44–49*   **23**

References and resources

Williams, E. M. and Shuard, H. *Primary Mathematics
Today* Third Edition (Chapters 7 and 30), Longman
Group Ltd 1982

E. J. Arnold *Hardwood 30 cm rule (cm and half-cm),
Clock face rubber stamps*

Nottingham Educational Supplies, *30 cm general
purpose rule (cm and
half-cm)*

Philip & Tacey Ltd, *Clock face rubber stamps,
Estimating and measuring square cm areas, Metre
measuring tape (cm and mm)*

Taskmaster Ltd, *Flexible 1 m measures (T373) cm and
half-cm*

Plastic measuring cylinders (marked in 100 ml)

Area 2

For the teacher

Finding the area of a triangle is tackled in a very practical way using examples where the triangle can be shown to have half the area of an enclosing rectangle.

The final section of the Teachers' Notes goes slightly beyond the work covered in the Pupils' Book in suggesting an approach for cases where the perpendicular height falls outside the base of a triangle.

Summary of the stages

1 Area of right-angled triangles
2 Area of other triangles

Vocabulary

Rectangle, diagonal, triangle, right angle, right-angled, base, perpendicular height.

Equipment and apparatus

Card or stiff paper, rulers, scissors, centimetre-squared paper, geoboards.

Working with the children

1 Area of right-angled triangles

Before rushing into the use of the formula $\frac{1}{2}(b \times h)$, it is essential that children gain practical experience of a triangle as half of a rectangle. The triangle which is half of a parallelogram will be introduced later. Cutting a card rectangle along the diagonal into halves will establish the reason for the '$\frac{1}{2}$' or '$\div 2$' in the final formula.

The diagrams in the Pupils' Book have been drawn at different angles. This is deliberate because it must not be suggested that area is altered by a change of orientation or that one side of a triangle must always be horizontal and the height vertical.

This would be a good time to remind children that *perpendicular* means 'at right angles to' and that although horizontal lines are perpendicular to vertical lines, two lines which are perpendicular (at right angles to each other) do not *have* to be horizontal or vertical. A simple diagram makes this clearer:

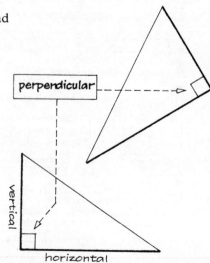

The 'square corner', ⌐, is

the conventional way of indicating a right angle.

Since 'length' and 'breadth' are inappropriate for a triangle, the terms *base* and *height* are introduced. In the first instance, the base and height of the triangle are shown as sides of the 'containing' rectangle. This helps to emphasise that the height is *at right angles to* the base; it is the *perpendicular* height. In some respects, the choice of the word *base* is unfortunate since it implies that this is the side on which the triangle appears to stand. It is probably better to think of the base as being the side to which a perpendicular is drawn from the *apex*. Some of the triangles are drawn upside down or 'base over apex' – an expression which has passed into the language to mean 'head over heels'.

In the first instance the formula for the area of a triangle is presented as (base × height) ÷ 2 or as $\frac{\text{base} \times \text{height}}{2}$ as these versions depict the 'order of operations' used in the preceding practical examples. That is: find the area of the containing rectangle first, then halve it to find the area of the triangle.

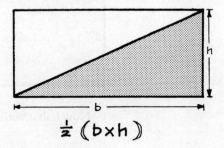

$\frac{1}{2}(b \times h)$

Other versions such as (½ base) × height or base × (½ height) should be left for possible later development.

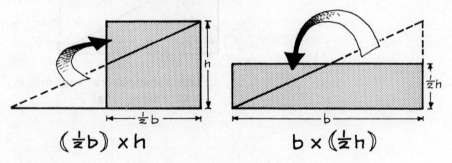

$(\frac{1}{2}b) \times h$ 　　　　 $b \times (\frac{1}{2}h)$

2   Area of other triangles

Again, it is essential that children go through the practical experience of cutting, turning and matching the pieces as shown in the diagram. A further useful activity is to give each child in a group one of a set of identical rectangles in paper or card. Each child chooses any point on one of the longer sides of his rectangle. This point becomes the apex of a triangle when it is joined to the ends of the side opposite:

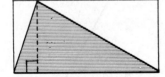 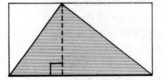 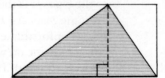 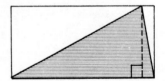

Although the resulting triangles are not the same shape – they look different – they all have the same area because each is half the area of one of the equal rectangles. Alternatively, as all these triangles have equal bases and equal perpendicular heights, the formula $\dfrac{\text{base} \times \text{height}}{2}$ will work out to the same in each case.

The examples in the Pupils' Book include some which are 'base over apex' but in all cases the base and height are chosen so that the perpendicular from the apex falls *within* the length of base.

What happens if the perpendicular height is *outside* the triangle? One way of dealing with this obtuse-angled triangle is to re-name the longest side (BC) as the base, then draw in and measure the perpendicular height to the new apex (A).

An alternative approach, which could be discussed with *some* children, uses the 'put and take' method on a geoboard as follows:

This rectangle has the same area as this oblique parallelogram.

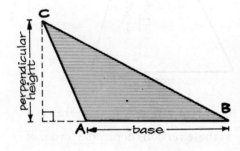

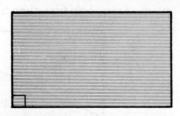 

Now halve both figures by drawing diagonals.

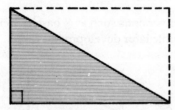 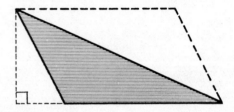

This triangle has the same area as   this triangle
(half of the rectangle)   (half of the parallelogram).

Pages from the Pupils' Book and Spiritmasters

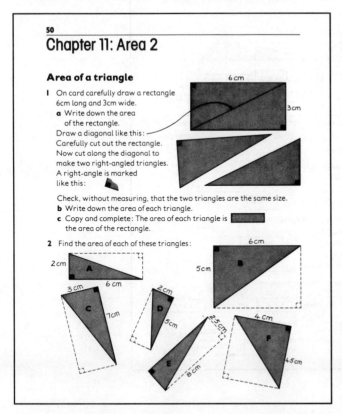

50

Chapter 11: Area 2

Area of a triangle

1 On card carefully draw a rectangle
   6cm long and 3cm wide.
   a Write down the area
      of the rectangle.
   Draw a diagonal like this:
   Carefully cut out the rectangle.
   Now cut along the diagonal to
   make two right-angled triangles.
   A right-angle is marked
   like this:

   Check, without measuring, that the two triangles are the same size.
   b Write down the area of each triangle.
   c Copy and complete: The area of each triangle is ▭
      the area of the rectangle.

2 Find the area of each of these triangles:

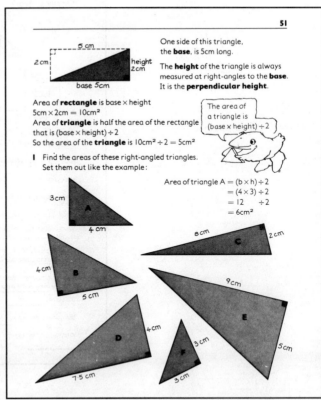

51

One side of this triangle,
the **base**, is 5cm long.

The **height** of the triangle is always
measured at right-angles to the **base**.
It is the **perpendicular height**.

Area of **rectangle** is base × height
5cm × 2cm = 10cm²
Area of **triangle** is half the area of the rectangle
that is (base × height) ÷ 2
So the area of the **triangle** is 10cm² ÷ 2 = 5cm²

The area of
a triangle is
(base × height) ÷ 2

1 Find the areas of these right-angled triangles.
   Set them out like the example:

Area of triangle A = (b × h) ÷ 2
            = (4 × 3) ÷ 2
            = 12   ÷ 2
            = 6cm²

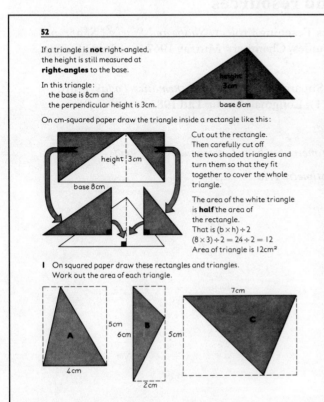

52

If a triangle is **not** right-angled,
the height is still measured at
right-angles to the base.

In this triangle:
   the base is 8cm and
   the perpendicular height is 3cm.

On cm-squared paper draw the triangle inside a rectangle like this:

Cut out the rectangle.
Then carefully cut off
the two shaded triangles and
turn them so that they fit
together to cover the whole
triangle.

The area of the white triangle
is **half** the area of
the rectangle.
That is (b × h) ÷ 2
(8 × 3) ÷ 2 = 24 ÷ 2 = 12
Area of triangle is 12cm²

1 On squared paper draw these rectangles and triangles.
   Work out the area of each triangle.

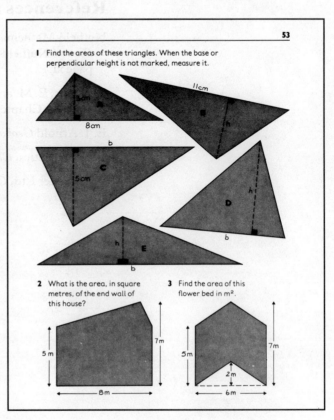

53

1 Find the areas of these triangles. When the base or
   perpendicular height is not marked, measure it.

2 What is the area, in square
   metres, of the end wall of
   this house?

3 Find the area of this
   flower bed in m².

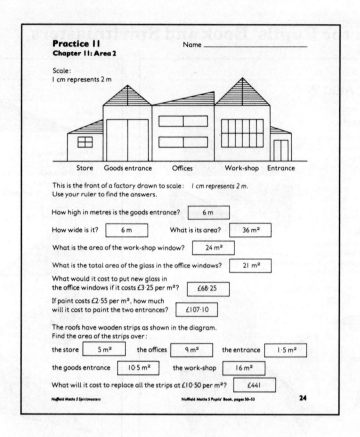

Practice 11
Chapter 11: Area 2

Name _____

Scale:
1 cm represents 2 m

Store   Goods entrance   Offices   Work-shop   Entrance

This is the front of a factory drawn to scale: *1 cm represents 2 m*.
Use your ruler to find the answers.

How high in metres is the goods entrance? | 6 m

How wide is it? | 6 m | What is its area? | 36 m²

What is the area of the work-shop window? | 24 m²

What is the total area of the glass in the office windows? | 21 m²

What would it cost to put new glass in
the office windows if it costs £3·25 per m²? | £68·25

If paint costs £2·55 per m², how much
will it cost to paint the two entrances? | £107·10

The roofs have wooden strips as shown in the diagram.
Find the area of the strips over:

the store | 5 m² | the offices | 9 m² | the entrance | 1·5 m²

the goods entrance | 10·5 m² | the work-shop | 16 m²

What will it cost to replace all the strips at £10·50 per m²? | £441

Nuffield Maths 5 Spiritmasters   *Nuffield Maths 5 Pupils' Book, pages 50–53*   **24**

References and resources

Nuffield Mathematics Teaching Project, *Shape and Size* ▽, *Shape and Size* ▽ Nuffield Guides, Chambers/Murray 1967 (See Introduction, page xi.)

Williams, E. M. and Shuard, H. *Primary Mathematics Today* Third Edition (Chapter 11), Longman Group Ltd 1982

E. J. Arnold *Geoboards*

Invicta Plastics, *Centimetre pin board, Area measuring grid*

Taskmaster Ltd, *Centimetre grid*

Co-ordinates

For the teacher

Emphasis is again put on the order of the pair of co-ordinates and the chapter begins with revision of this. A straight line graph is then drawn to introduce reading points on a graph and the chapter finishes with a look at simple transformations.

Summary of the stages

1 Revision of plotting points
2 A straight line graph
3 Plotting shapes
4 Simple transformations: a) translations, b) enlargements, c) reflections

Vocabulary

Lattice, isosceles, symmetry, axis, axes, horizontal, vertical, area, original, enlargements, reflection, translation.

Equipment and apparatus

Centimetre-squared paper, ruler, coloured crayons or pens.

Working with the children

1 Revision of plotting points
The order in which co-ordinates are listed (horizontal distance followed by vertical distance) in order to fix the position of a point, is an international convention. A visual reminder is given of this followed by points to be plotted which produce amusing pictures. The children should be encouraged to make up their own pictures and then list the co-ordinates for others to use. Plotting points should become second nature to the children. (*Nuffield Maths 5 Spiritmasters*, Grid 26.)

2 A straight line graph
The children are asked to list all the pairs of whole numbers which add up to 10. This is an opportunity to remind them of the commutativity of addition. For example (3, 7) and (7, 3) are two pairs of numbers which add up to 10. The diagram shows the straight line graph which results from plotting and joining the points.

Often additional information can be taken from a continuous line graph. This is called *interpolation* (from 'between points'). For instance, we can use the graph to complete pairs involving halves which add to ten. The scale of two squares to each number enables the child to write in the halves on each scale. They can then be used as in the diagram to complete the ordered pair given. For example, to complete $(4\frac{1}{2}, \Box)$ the broken line goes

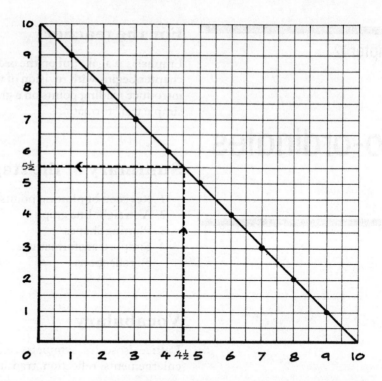

up from $4\frac{1}{2}$ to the straight line graph and then across to read $5\frac{1}{2}$ on the vertical axis, giving the pair $(4\frac{1}{2}, 5\frac{1}{2})$.

3   Plotting shapes

This exercise is revision of plotting points, naming shapes and marking in axes of symmetry. It should present little difficulty to the pupils but care must be taken in naming the shapes. For example, the shape marked **d** in the *Pupils' Book* and shown in this figure is, of course, an octagon but it is *not* a regular octagon. A regular octagon has its 8 sides and its 8 angles equal; this one does not. Four of its sides are 2 units long but the others are longer, as can be seen in the smaller diagram. Of all the regular polygons it is only possible to draw the square on a lattice with each vertex at the intersection of two lines. (*Nuffield Maths 5 Spiritmasters*, Grid 26.)

4   Simple transformations

All good pantomimes have a 'transformation scene' where Cinderella's kitchen becomes a ballroom or Aladdin's rags become expensive clothes. Mathematical transformations are similar in that they change shapes in a variety of ways. In this section some of the simpler transformations are introduced.

a)   Translations

A translation is the simplest of transformations as it is merely a 'sliding along' of the shape.

In this diagram the rectangle ABCD has been moved to a new position marked A′ B′ C′ D′. Each point has been moved 6 squares to the right and if we look at the co-ordinates we get:

Original set: A (2, 1)   B (2, 4)   C (6, 4)   D (6, 1)
New set: A (8, 1)   B (8, 4)   C (12, 4) D (12, 1)

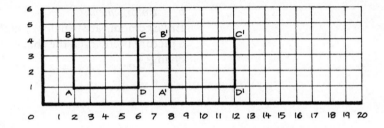

We have moved the shape 6 units to the right by adding 6 to the first number in each of the set of co-ordinates. In the *Pupils' Book* it is suggested that different colours are used for the new positions of lettered points rather than A′, B′, C′, etc.

We could move the shape to the left by *subtracting* the same from each of the first numbers in the set of co-ordinates. By adding the same to each of the second numbers in the set of co-ordinates, we move the figure up the page and by subtraction move it down the page.

For example, our original shape shaded on the lattice below has a set of co-ordinates: (7, 6), (9, 6), (8, 9)

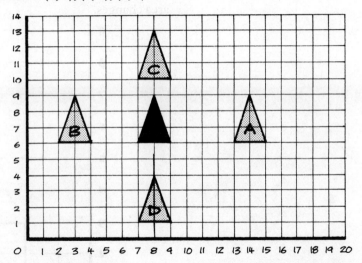

Add 6 to the first number in each:	(13, 6), (15, 6), (14, 9)	to give shape A 6 units to the right
Add 4 to the second number in each:	(7,10), (9,10), (8,13)	to give shape C 4 units up
Subtract 5 from the first number in each:	(2, 6), (4, 6), (3, 9)	to give shape B 5 units to the left
Subtract 5 from the second number in each:	(7, 1), (9, 1), (8, 4)	to give shape D 5 units down

In the last question of the exercise, a translation is made by adding to both numbers giving, in this case, a translation formed by a move of 4 to the right, combined with a move of 5 upwards.

The children are asked to confirm that although the *position* of a shape is altered by translation, its *area* remains unchanged. (*Nuffield Maths 5 Spiritmasters*, Grid 24.)

b) Enlargements

Enlargements are usually thought of as photographic enlargements. When a picture is enlarged the shape of the image remains unchanged but the area changes.

Mathematical enlargements are similar to this. Under an enlargement, a figure retains its shape (in mathematical terms its shape is *invariant*) but its area changes.

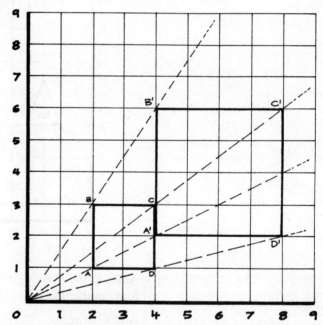

In this diagram the original shape is the square ABCD. Its area is 4 squares. Multiplying its co-ordinates by 2 gives a new set of points:

Original set   A (2,1),   B (2,3),   C (4,3),   D (4,1)
New set   A′ (4,2),   B′ (4,6),   C′ (8,6),   D′ (8,2)

This is shown on the diagram as the square A′B′C′D′.

It is still the same shape but its area is now 16 squares. If the lengths are multiplied by 2 the area is multiplied by 2^2 or 4. If the lengths are multiplied by 3, the area becomes 3^2 or 9 times as large, and so on.

In the diagram, dashed lines join each new point to its original and it can be seen that these lines all meet at the point (0,0). (Mathematically, this is called the *centre of enlargement*.)

The exercise for the pupils covers these ideas in simple steps. However, at this stage, no attempt should be made to formalise any conclusions the children may reach.

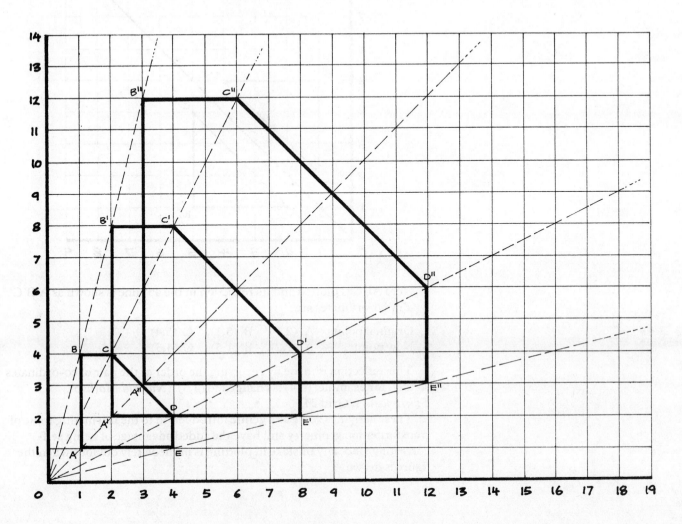

This diagram shows the finished exercise. Some children may be able to plot the points that a multiplication by, for example, $2\frac{1}{2}$ would give. These would still be on the broken lines. The graph can also be extended to give larger figures by extending the axes. Mathematically, multiplication by $\frac{1}{2}$, which would give a smaller figure than the original, is still called an enlargement.

This technique is useful for teachers wishing to make enlargements or reductions of diagrams and pictures without grappling with a pantograph. (*Nuffield Maths 5 Spiritmasters*, Grid 26.)

c) Reflections

Reflections are an extension of earlier work on symmetry. Any line of reflection acts as a mirror or axis of symmetry.

In this exercise the children look at a special reflection in a line at 45° to the horizontal axis.

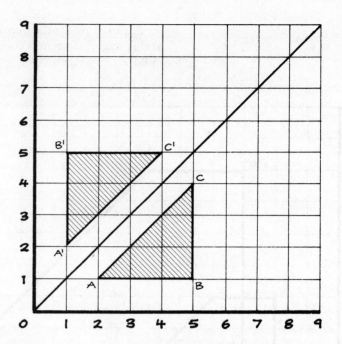

Here ABC is plotted and its reflection in the 45° line is shown as A'B'C'. The co-ordinates are:

Original points   A (2,1),   B (5,1),   C (5,4)
New points   A' (1,2),   B' (1,5),   C' (4,5)

This reflection is made by reversing the order of the pair of co-ordinates so that, for example, (5,4) is changed to (4,5). (*Nuffield Maths 5 Spiritmasters*, Grid 24)

These activities serve as a gentle introduction to the fascinating topic of transformation geometry and have the added advantage of being self-correcting since any mistake in plotting is immediately obvious when the figure is drawn.

Pages from the Pupils' Book and Spiritmasters

54

Chapter 12: Co-ordinates

Plotting points

I   Copy this lattice on to centimetre squared paper using one square on your paper for one square in the diagram.

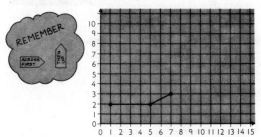

Mark and plot these points, joining them together as you plot them (the first three are drawn for you):
(1,2); (5,2); (7,3); (9,3); (10,4); (12,4); (14,5); (12,6); (10,6); (9,7); (7,7); (5,8); (1,8); (4,7); (4,3); (1,2).

2   Make another lattice on squared paper like the last one.
Mark and plot these points joining them together as you plot them.
(2,1); (2,6); (4,8); (6,8); (8,7); (10,8); (11,7); (12,4); (11,2); (10,2); (11,4); (10,5); (9,4); (9,1); (8,1); (8,4); (6,3); (3,4); (3,1); (2,1).

Then join (2,6) to (1,5),
and join (9,5) to (7,5) to (9,7).
Put an eye at (10,6).

3   Make up a picture of your own. List the points in order and give it to your friend for him to draw.

55

A straight line graph

$3 + 7 = 10$
$8 + 2 = 10$

These are two pairs of whole numbers which add up to 10.
They can be written as:
(3,7); (8,2)

I   a   Make a list of all the pairs of numbers which add up to 10 —don't forget (0,10).

   b   Copy this lattice onto centimetre squared paper using a square centimetre for each square in the diagram. Notice that each number takes two squares

   c   Plot the pairs of numbers you listed. Two are done for you.

   d   The points you have marked should all lie on a straight line. Draw the line.
The line is the graph of pairs of numbers which add up to ten.

   e   As two squares were used for one whole number when we marked our lattice, the squares between are used for halves. Write them on your lattice as in this diagram.

   f   Use the graph to complete these pairs of numbers which add up to 10:

$(7\frac{1}{2}, \blacksquare)$; $(\blacksquare, 3\frac{1}{2})$; $(1\frac{1}{2}, \blacksquare)$; $(5\frac{1}{2}, \blacksquare)$; $(\blacksquare, 9\frac{1}{2})$;

$(8\frac{1}{2}, \blacksquare)$; $(\blacksquare, 4\frac{1}{2})$.

56

Plotting shapes

I   a   Copy the lattice in the diagram on to centimetre squared paper.

   b   Mark and join together, in order, the points:
(1,1); (5,1); (3,7); (1,1)

The shape drawn should be the isosceles triangle marked **a** on the diagram.

This shape has one axis of symmetry and it has been drawn on the figure with a dashed line.

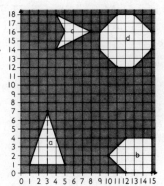

2   Copy the other three shapes onto the lattice making a list of the points plotted for each figure.
With a dashed line, mark in any axes of symmetry.

3   Plot these points joining them together in order.
Name the shape drawn and mark in any axes of symmetry.

   a   (1,13); (4,13); (4,10); (1,10); (1,13).
   b   (6,0); (9,0); (9,4); (6,4); (6,0).
   c   (12,7); (12,11); (15,10); (15,6); (12,7).
   d   (5,12); (6,14); (8,14); (9,12); (8,10); (6,10); (5,12).
   e   (5,6); (11,6); (9,9); (5,9); (5,6).

57

Translations

I   a   Draw a lattice on centimetre squared paper. Number the horizontal → axis from 0 to 10 and the vertical ↑ axis from 0–10.

   b   Plot and join the points in order:
A (1,1); B (1,4); C (4,4); D (4,1); A (1,1).

   c   What shape have you drawn?
   d   What is the area of the shape?

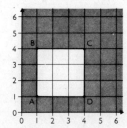

2   Adding 4 to the first number in each pair gives us a new set of points:
Original set: A (1,1); B (1,4); C (4,4); D (4,1); A (1,1).
New set: A (5,1); B (5,4); C (8,4); D (8,1); A (5,1).

   a   Mark and join up the new set of points in a different colour.
   b   What shape have you drawn?
   c   What is its area?
   d   What has happened to the original shape?

3   a   Make a new set of points by adding 5 to the second number in each pair of the **original** set.
   b   Mark and join up the new set of points in a different colour.
   c   What shape have you drawn?
   d   What is its area?
   e   What has happened to it?

4   a   Make a new set of points by adding 4 to the first number and by adding 5 to the second number in each pair of the **original** set.
   b   Mark and join up the new set of points in a different colour.
   c   What shape have you drawn?
   d   What is its area?
   e   What has happened to it?

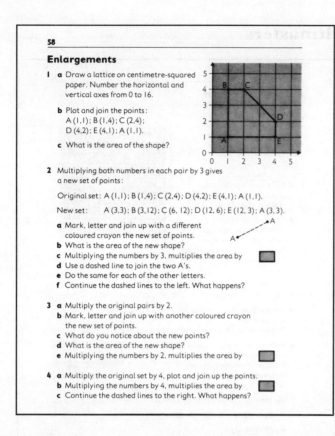

58

Enlargements

1 a Draw a lattice on centimetre-squared paper. Number the horizontal and vertical axes from 0 to 16.

b Plot and join the points:
A (1,1); B (1,4); C (2,4); D (4,2); E (4,1); A (1,1).

c What is the area of the shape?

2 Multiplying both numbers in each pair by 3 gives a new set of points:

Original set: A (1,1); B (1,4); C (2,4); D (4,2); E (4,1); A (1,1).

New set: A (3,3); B (3,12); C (6,12); D (12,6); E (12,3); A (3,3).

a Mark, letter and join up with a different coloured crayon the new set of points.

b What is the area of the new shape?

c Multiplying the numbers by 3, multiplies the area by

d Use a dashed line to join the two A's.

e Do the same for each of the other letters.

f Continue the dashed lines to the left. What happens?

3 a Multiply the original pairs by 2.

b Mark, letter and join up with another coloured crayon the new set of points.

c What do you notice about the new points?

d What is the area of the new shape?

e Multiplying the numbers by 2, multiplies the area by

4 a Multiply the original set by 4, plot and join up the points.

b Multiplying the numbers by 4, multiplies the area by

c Continue the dashed lines to the right. What happens?

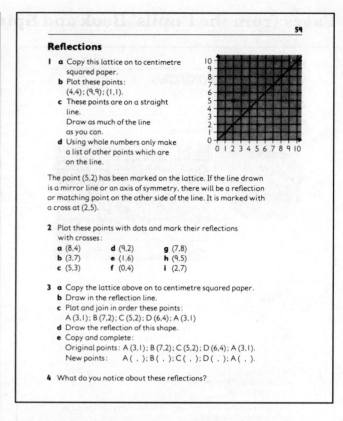

59

Reflections

1 a Copy this lattice on to centimetre squared paper.

b Plot these points:
(4,4); (9,9); (1,1).

c These points are on a straight line.
Draw as much of the line as you can.

d Using whole numbers only make a list of other points which are on the line.

The point (5,2) has been marked on the lattice. If the line drawn is a mirror line or an axis of symmetry, there will be a reflection or matching point on the other side of the line. It is marked with a cross at (2,5).

2 Plot these points with dots and mark their reflections with crosses:

a (8,4)	**d** (9,2)	**g** (7,8)
b (3,7)	**e** (1,6)	**h** (9,5)
c (5,3)	**f** (0,4)	**i** (2,7)

3 a Copy the lattice above on to centimetre squared paper.

b Draw in the reflection line.

c Plot and join in order these points:
A (3,1); B (7,2); C (5,2); D (6,4); A (3,1)

d Draw the reflection of this shape.

e Copy and complete:
Original points: A (3,1); B (7,2); C (5,2); D (6,4); A (3,1).
New points: A (,); B (,); C (,); D (,); A (,).

4 What do you notice about these reflections?

References and resources

Nuffield Maths Teaching Project, *Graphs leading to Algebra* ②, Nuffield Teaching Guides, Chambers/Murray 1969 (See Introduction page xi.)

Williams, E. M. and Shuard, H. *Primary Mathematics Today* Third Edition (Chapter 12), Longman Group 1982

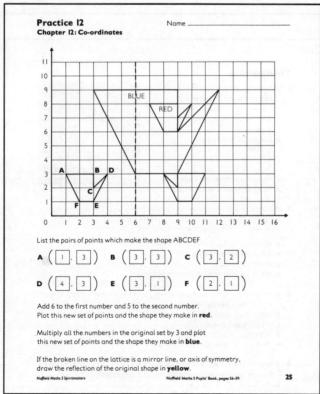

Practice 12
Chapter 12: Co-ordinates   Name _____

List the pairs of points which make the shape ABCDEF

A (1 , 3)   **B** (3 , 3)   **C** (3 , 2)

D (4 , 3)   **E** (3 , 1)   **F** (2 , 1)

Add 6 to the first number and 5 to the second number.
Plot this new set of points and the shape they make in **red**.

Multiply all the numbers in the original set by 3 and plot this new set of points and the shape they make in **blue**.

If the broken line on the lattice is a mirror line, or axis of symmetry, draw the reflection of the original shape in **yellow**.

Nuffield Maths 5 Spiritmasters   Nuffield Maths 5 Pupils' Book, pages 54–59   **25**

Weight

For the teacher

After some exercises arising from the nett and gross weights of packages, this chapter strengthens the link between the place value and the recording of weights in decimal notation. Much of this work runs parallel to previous activities with money, length and volume.

Summary of the stages

1 Nett weight and gross weight
2 Recording on the abacus – decimal notation
3 Multiplication and division of weight

Vocabulary

Nett, gross, kilogram (kg), gram (g), 'hecto', estimate, approximate.

Equipment and apparatus

Scales with dials, 4 or 5 spiked abacus, labels, packets, tins etc., showing nett weights of foods and other products.

Working with the children

1 Nett weight and gross weight
Discussion of the nett and gross weights of packaged foods, soap powders, etc. leads to revision of addition and subtraction of weights and the accurate reading of scales. The exercises in the *Pupils' Book* should be supplemented by further examples from the children's environment. It is advisable to remind children that manufacturers are notoriously careless about the use of abbreviations for metric measurements and that the correct abbreviations are kg for kilogram(s) and g for gram(s).
(*Nuffield Maths 5 Spiritmasters*, Grid 12.)

2 Recording on the abacus – decimal notation
Earlier work has made the gradual progression in recording weights in kilograms and grams (and capacities in litres and millilitres) as follows:

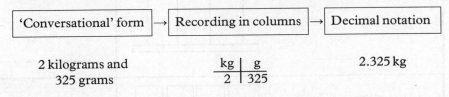

| 'Conversational' form | → | Recording in columns | → | Decimal notation |

| | kg | g | |
| 2 kilograms and 325 grams | 2 | 325 | 2.325 kg |

This section reinforces the last step in the progression and helps to establish the relationship between kilograms and parts of a kilogram.

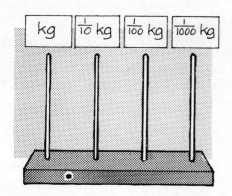

The decimal point must be positioned on the base of the abacus so that there are three spikes to the right of the point. The labelling of the abacus columns can be changed when appropriate:

from | 1 kg | $\frac{1}{10} \text{ kg}$ | $\frac{1}{100} \text{ kg}$ | $\frac{1}{1000} \text{ kg}$

to | kg | 0.1 kg | 0.01 kg | 0.001 kg

(*Nuffield Maths 5 Spiritmasters*, Grid 30.)

Additional practice at both abacus recording and estimation can be given by providing copies of sheets like this: (*Nuffield Maths 5 Spiritmasters*, Grid 23.)

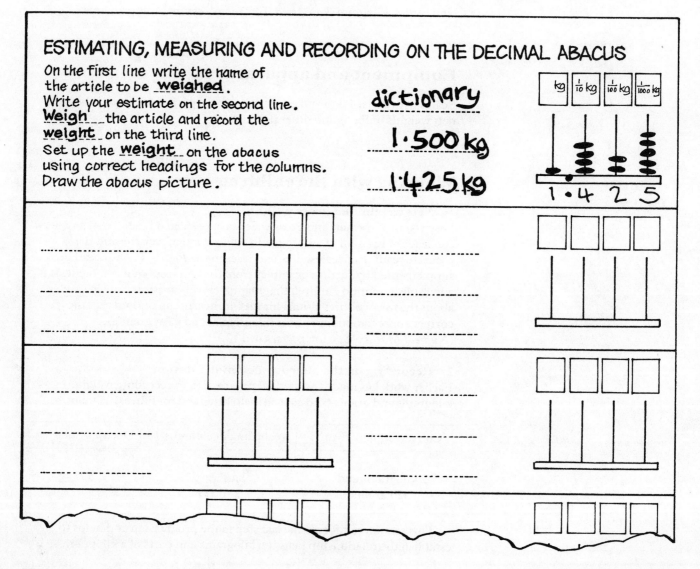

Once the relationship between parts of a kilogram has been established, the comparison and ordering of weights expressed in different forms can be undertaken. This is achieved more easily if all the weights are first changed to decimal notation.

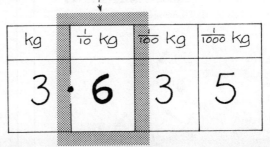

		kg		*kg*
30 g	⟶	0.030		→0.900
$\frac{9}{10}$	⟶	0.900		0.400
0.060 kg	⟶	0.060		→0.060
50 g	⟶	0.050		→0.050
4 'hectos'	⟶	0.400		0.030

When reading off to the nearest kilogram, it is the $\frac{1}{10}$ kg or 0.1 kg digit which is important.

kg	$\frac{1}{10}$ kg	$\frac{1}{100}$ kg	$\frac{1}{1000}$ kg
3 ·	6	3	5

The other digits in the columns to the right can be ignored. (See Chapter 10 on 'Rounding off'.)

3.635 kg ≃ 4 kg

3   Multiplication and division of weight

The layout for multiplication of kg and g by a single digit follows the same pattern as that used for multiplication of m and cm (Chapter 8: Length 1) except that this time there are three decimal places. Whether the 'partial products' or 'telescoped' format is used will depend on the child's level of ability and confidence. In either case, 'keeping the points in line' must be emphasised.

Again, it is important to establish the habit of 'thinking round the problem' and recording an estimate first. (See reference to 'Think before ink' on page 47 of this handbook.)

Division of kg and g by a single digit can be done either by working in kilograms (keeping the points in line) or by working in grams. These alternatives are discussed in Chapter 8: Length 1, page 44.

Pages from the Pupils' Book and Spiritmasters

60

Chapter 13 : Weight

Nett weight and gross weight

This tin of peas weighs 520g but the label says 425g.
This means that the peas inside the tin
weigh 425g—this is the nett weight.
The peas and the tin together weigh 520g—
this is called the gross weight.

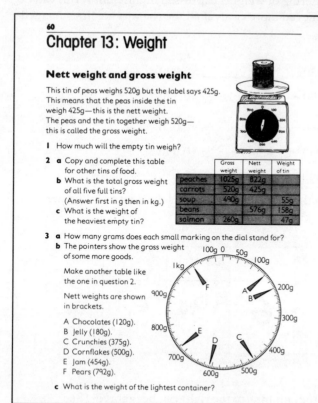

1 How much will the empty tin weigh?

2 a Copy and complete this table
for other tins of food.

b What is the total gross weight
of all five full tins?
(Answer first in g then in kg.)

c What is the weight of
the heaviest empty tin?

	Gross weight	Nett weight	Weight of tin
peaches	1025g	822g	
carrots	520g	425g	
soup	490g		55g
beans		576g	158g
salmon	260g		47g

3 a How many grams does each small marking on the dial stand for?

b The pointers show the gross weight
of some more goods.

Make another table like
the one in question 2.

Nett weights are shown
in brackets.

A Chocolates (120g).
B Jelly (180g).
C Crunchies (375g).
D Cornflakes (500g).
E Jam (454g).
F Pears (792g).

c What is the weight of the lightest container?

61

$1000g = 1$ kilogram.

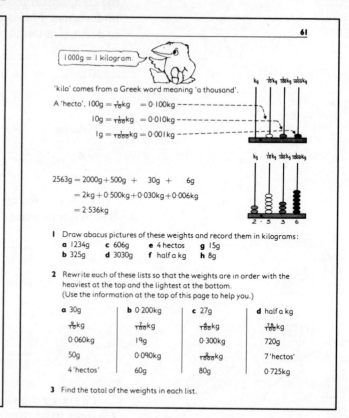

'kilo' comes from a Greek word meaning 'a thousand'.

A 'hecto', $100g = \frac{1}{10}kg = 0.100kg$ -------

$10g = \frac{1}{100}kg = 0.010kg$ -------

$1g = \frac{1}{1000}kg = 0.001kg$ -------

$2563g = 2000g + 500g + 30g + 6g$

$= 2kg + 0.500kg + 0.030kg + 0.006kg$

$= 2.536kg$

1 Draw abacus pictures of these weights and record them in kilograms :

a 1234g **c** 606g **e** 4 hectos **g** 15g

b 325g **d** 3030g **f** half a kg **h** 8g

2 Rewrite each of these lists so that the weights are in order with the
heaviest at the top and the lightest at the bottom.
(Use the information at the top of this page to help you.)

a	b	c	d
30g	0.200kg	27g	half a kg
$\frac{8}{10}$kg	$\frac{7}{100}$kg	$\frac{1}{100}$kg	$\frac{75}{100}$kg
0.060kg	19g	0.300kg	720g
50g	0.090kg	$\frac{2}{1000}$kg	7 'hectos'
4 'hectos'	60g	80g	0.725kg

3 Find the total of the weights in each list.

62

1

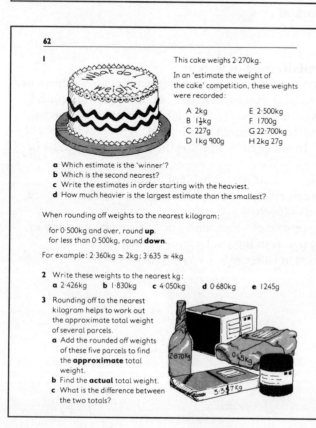

This cake weighs 2.270kg.

In an 'estimate the weight of
the cake' competition, these weights
were recorded :

A 2kg   E 2.500kg
B 1½kg   F 1700g
C 227g   G 22.700kg
D 1kg 900g H 2kg 27g

a Which estimate is the 'winner'?
b Which is the second nearest?
c Write the estimates in order starting with the heaviest.
d How much heavier is the largest estimate than the smallest?

When rounding off weights to the nearest kilogram :

for 0.500kg and over, round **up**.
for less than 0.500kg, round **down**.

For example : $2.360kg \simeq 2kg$; $3.635 \simeq 4kg$

2 Write these weights to the nearest kg :
a 2.426kg **b** 1.830kg **c** 4.050kg **d** 0.680kg **e** 1245g

3 Rounding off to the nearest
kilogram helps to work out
the approximate total weight
of several parcels.
a Add the rounded off weights
of these five parcels to find
the **approximate** total
weight.
b Find the **actual** total weight.
c What is the difference between
the two totals?

63

Multiplication of kg and g

The gross weight of
a tin of plums is 1.135kg.
To find the total weight
of 5 tins, multiply
1.135kg by 5.
(Estimate : $1.135kg \times 5$
$\simeq 5.500kg$)

$$\begin{array}{r} 1.135kg \\ \times\ 5 \\ \hline .025 \\ .150 \\ .500 \\ 5.000 \\ \hline 5.675kg \end{array}$$
$(.005 \times 5)$
$(.030 \times 5)$
$(.100 \times 5)$
(1.000×5)

or
$$\begin{array}{r} 1.135kg \\ \times\ 5 \\ \hline 5.675kg \\ {}_{1\ 2} \end{array}$$

The decimal
points
must be in line.

1 The nett weight of the plums in each tin is 937 grams.
What is the total nett weight of the plums from 5 tins?

2 Find the difference between the total gross weight and the total nett
weight of 7 tins of pears if 1 tin has a gross weight of 1.025 and a nett
weight of 792g. (Estimate first.)

Division of kg and g

If 4 bottles of lemonade weigh 2.328kg what does 1 bottle weigh?
(Estimate : $2.400 \div 4 = 0.600$, so $2.328kg \div 4$ is just under 0.600)

Working in kilograms : or Working in grams :

$$\begin{array}{r} 0.582kg \\ 4\overline{)2.328kg} \\ -2.000 \\ \hline 0.328 \\ -0.320 \\ \hline 0.008 \\ -0.008 \\ \hline 0.582kg \end{array}$$
0.500×4
0.080×4
0.002×4

$582g = 0.582kg$
$$\begin{array}{r} 4\overline{)2328g} \\ -2000 \\ \hline 328 \\ -320 \\ \hline 8 \\ -8 \\ \hline 582g \end{array}$$
500×4
80×4
2×4

3 Find the weight of 1 packet of soap-powder if the total weight of 6
packets is 13.860kg.

4 5 tins of tomatoes weigh 3.695kg and 7 tins of beans weigh 5.075kg.
Which is heavier, a tin of tomatoes or a tin of beans, and by how much?

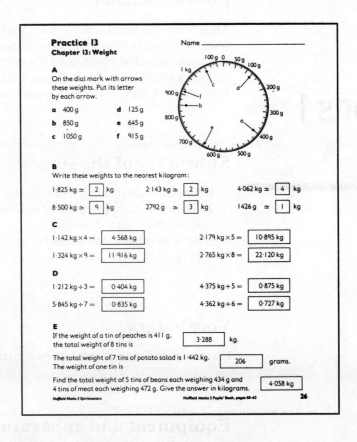

Practice 13
Chapter 13: Weight

Name _____

A
On the dial mark with arrows
these weights. Put its letter
by each arrow.

a 400 g   d 125 g

b 850 g   e 645 g

c 1050 g   f 915 g

B
Write these weights to the nearest kilogram:

1·825 kg ≃ ⟦2⟧ kg   2·143 kg ≃ ⟦2⟧ kg   4·062 kg ≃ ⟦4⟧ kg

8·500 kg ≃ ⟦9⟧ kg   2792 g ≃ ⟦3⟧ kg   1426 g ≃ ⟦1⟧ kg

C

1·142 kg × 4 = ⟦4·568 kg⟧   2·179 kg × 5 = ⟦10·895 kg⟧

1·324 kg × 9 = ⟦11·916 kg⟧   2·765 kg × 8 = ⟦22·120 kg⟧

D

1·212 kg ÷ 3 = ⟦0·404 kg⟧   4·375 kg ÷ 5 = ⟦0·875 kg⟧

5·845 kg ÷ 7 = ⟦0·835 kg⟧   4·362 kg ÷ 6 = ⟦0·727 kg⟧

E

If the weight of a tin of peaches is 411 g,
the total weight of 8 tins is   ⟦3·288⟧ kg.

The total weight of 7 tins of potato salad is 1·442 kg.
The weight of one tin is   ⟦206⟧ grams.

Find the total weight of 5 tins of beans each weighing 434 g and
4 tins of meat each weighing 472 g. Give the answer in kilograms.   ⟦4·058 kg⟧

Nuffield Maths 5 Spiritmasters   *Nuffield Maths 5 Pupils' Book, pages 60–63*   **26**

References and resources

Arnold, E. J. *Abacus board and tablets, Harex Abacus*

E.S.A. *5 kg Balance (5 kg in 20 g's), 1000 g Balance (1 kg in 10 g's)*

Invicta Plastics, *Abacus and tablets*

Nicolas Burdett *Compact Scale (5 kg in 50 g's), Compression Scale (10 kg in 50 g calibrations), Flat Pan Scale (5 kg in 20 g's)*

Osmiroid, *Abacent*

Philip & Tacey Ltd, *Bantam Weight Scales (750 g in 5 g's), Metway General Purpose Scales (5 kg in 25 g's)*

Taskmaster Ltd, *Stowaway Scales (5 kg in 25 g's) Abacus boards*

Triman Classmate, *Decimal abacus*

Chapter 14

Fractions 1

For the teacher

After brief revision of fraction families, multiplication and division of a fraction by one is dealt with at some length. Since this is possibly the most important aspect of fraction work it should not be rushed. Addition and subtraction go hand in hand through the various stages to the point where fractions with different denominators are being dealt with.

Summary of the stages

1 Multiplication and division by one
2 Addition and subtraction of fractions
   a) with the same denominator
   b) with related denominators
   c) with different denominators

Vocabulary

Fraction, family, numerator, common denominator.

Equipment and apparatus

Squared paper, crayons.

Working with the children

1 Multiplication and division by one

As nearly all fraction work depends upon an ability to produce equivalent fractions this is again emphasised in the first stage. That multiplication by one leaves a number unchanged is readily accepted by children and the exercise on this should cause few difficulties. Division by one is a little harder and it may be necessary to look again at what is meant by division.

$12 \div 3$ may be read as 'How many threes are there in 12?'

It follows then that:

$7 \div 1$ may be read as 'How many ones are there in 7?'

The more children read division in this way the better. How many adults might give the answer 2 to the question $4 \div \frac{1}{2}$? Read it as 'How many halves in 4?' and the answer 8 is obvious.

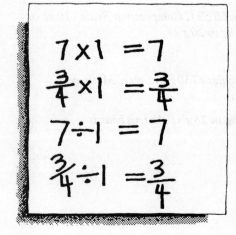

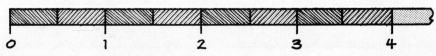

From the exercise a rule is formulated:

> The value of a fraction is unchanged if we multiply or divide the *numerator* (number above the line) and the *denominator* (number below the line) by the same number.

Practice at this follows and the children should be encouraged to think along these lines:

'The denominator has been multiplied by 4 so I must multiply the numerator by 4.'

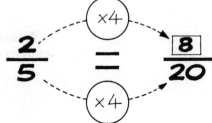

2 Addition and subtraction of fractions

a) *with the same denominator*
When the denominators of the fractions to be added or subtracted are the same, little difficulty arises provided the children have had practical experience of handling fractions. The folding of a piece of paper into eighths to answer $\frac{3}{8} + \frac{2}{8}$ is not a wasted exercise. The shading in of $\frac{3}{8}$ and $\frac{2}{8}$ shows that together

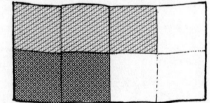

$$\frac{3}{8} + \frac{2}{8} = \frac{5}{8}$$

It is practical work like this and the use of fraction boards (See *Nuffield Maths 4 Teachers' Handbook*, page 99) which prevents the mistake

$$\frac{3}{8} + \frac{2}{8} = \frac{(3+2)}{(8+8)} = \frac{5}{16}$$

Sound also helps: 'Three *eighths* and two *eighths* make five *eighths*.'

b) *with related denominators*
The next set of examples are those where one of the fractions needs to be changed to make their denominators the same. Here is an example laid out with suggested thinking:

'I can't add fifths to tenths – they are different things, but 5 is a factor of 10 so I can change fifths to tenths.'

'I have multiplied 5 by 2, so I must multiply 3 by 2.

'6 tenths plus 3 tenths is 9 tenths.'

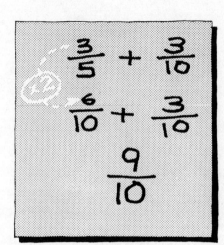

Writing $\frac{6}{10} + \frac{3}{10}$ in the second line is preferred to $\frac{6+3}{10}$. Many children have great difficulty in seeing that the $\frac{6+3}{10}$ form is in fact $\frac{6}{10} + \frac{3}{10}$.

c)   with different denominators

For example, $\dfrac{1}{2} + \dfrac{1}{3}$. Here we need to find a common denominator by

looking at the families of $\dfrac{1}{2}$ and $\dfrac{1}{3}$

$$\frac{1}{2} = \left\{ \begin{array}{c} \dfrac{1}{2}, \dfrac{2}{4}, \dfrac{3}{6}, \dfrac{4}{8}, \dfrac{5}{10} \cdots \end{array} \right\}$$

$$\frac{1}{3} = \left\{ \begin{array}{c} \dfrac{1}{3}, \dfrac{2}{6} \cdots \cdots \end{array} \right\}$$

We need go no further with the family of $\dfrac{1}{3}$ as we have found our common denominator

$$\frac{1}{2} + \frac{1}{3}$$

$$= \frac{3}{6} + \frac{2}{6}$$

$$= \frac{5}{6}$$

With simple fractions such as those in the exercise there should be no need to write more than 5 or 6 members of the first family to give the common denominator.

For example $\dfrac{9}{10} - \dfrac{3}{4}$

$$\frac{9}{10} = \left\{ \begin{array}{c} \dfrac{18}{20}, \dfrac{27}{30}, \dfrac{36}{40}, \dfrac{45}{50}, \dfrac{54}{60} \cdots \cdots \end{array} \right\}$$

$$= \frac{18}{20} - \frac{15}{20} \qquad \frac{3}{4} = \left\{ \begin{array}{c} \dfrac{6}{8}, \dfrac{9}{12}, \dfrac{12}{16}, \dfrac{15}{20} \cdots \cdots \end{array} \right\}$$

$$= \frac{3}{20}$$

Pages from the Pupils' Book and Spiritmasters

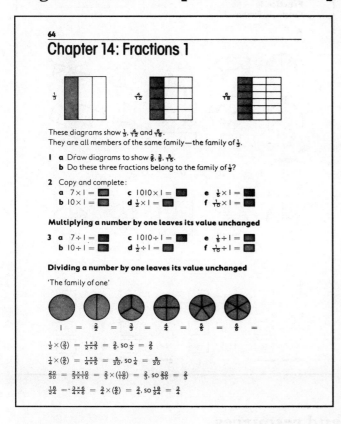

64

Chapter 14: Fractions 1

These diagrams show $\frac{1}{3}$, $\frac{4}{12}$ and $\frac{6}{18}$.
They are all members of the same family—the family of $\frac{1}{3}$.

1 a Draw diagrams to show $\frac{2}{6}$, $\frac{3}{8}$, $\frac{1}{15}$.
b Do these three fractions belong to the family of $\frac{1}{3}$?

2 Copy and complete:
a $7\times1=$ **c** $1010\times1=$ **e** $\frac{1}{5}\times1=$
b $10\times1=$ **d** $\frac{1}{2}\times1=$ **f** $\frac{1}{10}\times1=$

Multiplying a number by one leaves its value unchanged

3 a $7\div1=$ **c** $1010\div1=$ **e** $\frac{1}{5}\div1=$
b $10\div1=$ **d** $\frac{1}{2}\div1=$ **f** $\frac{1}{10}\div1=$

Dividing a number by one leaves its value unchanged

'The family of one'

$1 = \frac{2}{2} = \frac{3}{3} = \frac{4}{4} = \frac{5}{5} = \frac{6}{6} =$

$\frac{1}{2}\times(\frac{3}{3}) = \frac{1\times3}{2\times3} = \frac{3}{6}$, so $\frac{1}{2} = \frac{3}{6}$

$\frac{1}{4}\times(\frac{5}{5}) = \frac{1\times5}{4\times5} = \frac{5}{20}$, so $\frac{1}{4} = \frac{5}{20}$

$\frac{20}{30} = \frac{2\times10}{3\times10} = \frac{2}{3}\times(\frac{10}{10}) = \frac{2}{3}$, so $\frac{20}{30} = \frac{2}{3}$

$\frac{18}{24} = \frac{3\times6}{4\times6} = \frac{3}{4}\times(\frac{6}{6}) = \frac{3}{4}$, so $\frac{18}{24} = \frac{3}{4}$

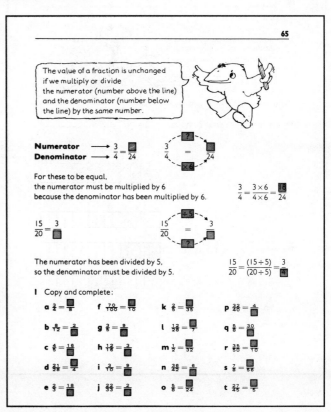

65

The value of a fraction is unchanged if we multiply or divide the numerator (number above the line) and the denominator (number below the line) by the same number.

Numerator $\longrightarrow$ $\frac{3}{4} = \frac{\square}{24}$
Denominator $\longrightarrow$

For these to be equal, the numerator must be multiplied by 6 because the denominator has been multiplied by 6.

$\frac{3}{4} = \frac{3\times6}{4\times6} = \frac{\square}{24}$

$\frac{15}{20} = \frac{3}{\square}$

The numerator has been divided by 5, so the denominator must be divided by 5.

$\frac{15}{20} = \frac{(15\div5)}{(20\div5)} = \frac{3}{\square}$

1 Copy and complete:
a $\frac{3}{4} = \frac{\square}{8}$ **f** $\frac{70}{100} = \frac{\square}{10}$ **k** $\frac{2}{5} = \frac{\square}{35}$ **p** $\frac{32}{40} = \frac{4}{\square}$
b $\frac{8}{12} = \frac{2}{\square}$ **g** $\frac{3}{5} = \frac{9}{\square}$ **l** $\frac{12}{28} = \frac{\square}{7}$ **q** $\frac{5}{8} = \frac{30}{\square}$
c $\frac{4}{5} = \frac{16}{\square}$ **h** $\frac{12}{18} = \frac{3}{\square}$ **m** $\frac{1}{2} = \frac{\square}{32}$ **r** $\frac{35}{50} = \frac{\square}{10}$
d $\frac{21}{28} = \frac{\square}{4}$ **i** $\frac{7}{10} = \frac{9}{\square}$ **n** $\frac{36}{42} = \frac{6}{\square}$ **s** $\frac{7}{8} = \frac{\square}{56}$
e $\frac{2}{3} = \frac{18}{\square}$ **j** $\frac{22}{33} = \frac{2}{\square}$ **o** $\frac{5}{8} = \frac{\square}{24}$ **t** $\frac{27}{45} = \frac{\square}{5}$

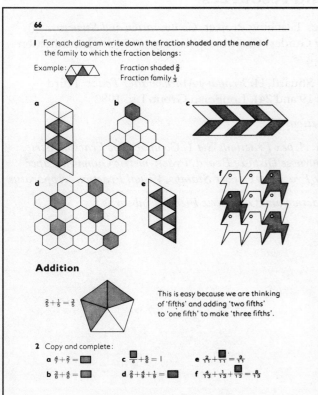

66

1 For each diagram write down the fraction shaded and the name of the family to which the fraction belongs:

Example: Fraction shaded $\frac{2}{6}$
Fraction family $\frac{1}{3}$

a b c
d e f

Addition

$\frac{2}{5} + \frac{1}{5} = \frac{3}{5}$

This is easy because we are thinking of 'fifths' and adding 'two fifths' to 'one fifth' to make 'three fifths'.

2 Copy and complete:
a $\frac{4}{7} + \frac{2}{7} =$ **c** $\frac{3}{8} + \frac{5}{8} = 1$ **e** $\frac{2}{11} + \frac{7}{11} = \frac{9}{11}$
b $\frac{3}{8} + \frac{4}{8} =$ **d** $\frac{2}{9} + \frac{4}{9} + \frac{1}{9} =$ **f** $\frac{4}{13} + \frac{1}{13} + \frac{3}{13} = \frac{8}{13}$

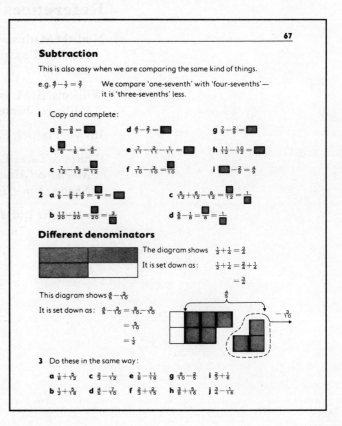

67

Subtraction

This is also easy when we are comparing the same kind of things.

e.g. $\frac{4}{7} - \frac{1}{7} = \frac{3}{7}$ We compare 'one-seventh' with 'four-sevenths'— it is 'three-sevenths' less.

1 Copy and complete:
a $\frac{5}{8} - \frac{3}{8} =$ **d** $\frac{4}{7} - \frac{2}{7} =$ **g** $\frac{7}{9} - \frac{2}{9} =$
b $\frac{\square}{6} - \frac{1}{6} = \frac{4}{6}$ **e** $\frac{2}{11} - \frac{2}{11} - \frac{1}{11} =$ **h** $\frac{11}{12} - \frac{10}{12} =$
c $\frac{7}{12} - \frac{5}{12} = \frac{\square}{12}$ **f** $\frac{7}{10} - \frac{3}{10} = \frac{\square}{10}$ **i** $\frac{\square}{9} - \frac{2}{9} = \frac{4}{9}$

2 a $\frac{7}{9} - \frac{2}{9} + \frac{3}{9} = \frac{\square}{9} =$ **c** $\frac{5}{12} + \frac{6}{12} - \frac{2}{12} = \frac{\square}{12} = \square$
b $\frac{17}{20} - \frac{11}{20} = \frac{\square}{20} = \frac{\square}{\square}$ **d** $\frac{5}{8} - \frac{1}{8} = \frac{\square}{8} = \frac{\square}{\square}$

Different denominators

The diagram shows $\frac{1}{2} + \frac{1}{4} = \frac{3}{4}$
It is set down as: $\frac{1}{2} + \frac{1}{4} = \frac{2}{4} + \frac{1}{4}$
$= \frac{3}{4}$

This diagram shows $\frac{4}{5} - \frac{3}{10}$
It is set down as: $\frac{4}{5} - \frac{3}{10} = \frac{8}{10} - \frac{3}{10}$
$= \frac{5}{10}$
$= \frac{1}{2}$

3 Do these in the same way:
a $\frac{1}{8} + \frac{5}{12}$ **c** $\frac{2}{3} - \frac{1}{12}$ **e** $\frac{7}{8} - \frac{11}{16}$ **g** $\frac{9}{10} - \frac{2}{5}$ **i** $\frac{2}{3} + \frac{1}{6}$
b $\frac{1}{2} + \frac{5}{16}$ **d** $\frac{4}{5} - \frac{7}{10}$ **f** $\frac{2}{3} + \frac{2}{15}$ **h** $\frac{3}{8} + \frac{5}{16}$ **j** $\frac{3}{4} - \frac{1}{16}$

81

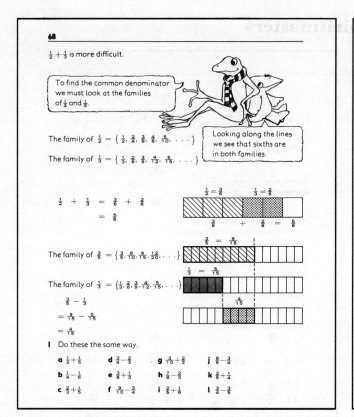

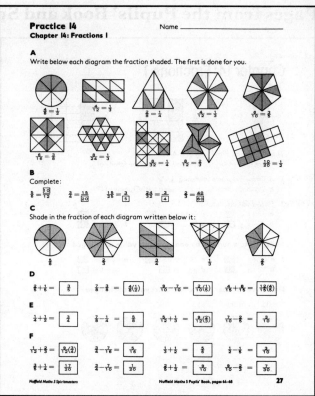

References and resources

Nuffield Mathematics Teaching Project, *Computation and Structure* ⑤ Nuffield Teaching Guides, Chambers/Murray 1967 (See Introduction page xi.)

Williams, E. M. and Shuard, H. *Primary Mathematics Today* Third Edition (Chapters 19 and 24), Longman Group Ltd 1982

Invicta Plastics, *Fractions kit*

Philips & Tacey Ltd, *Aspex Fractions Set 1, Comparative Fraction Strips, Magna-cel Multipurpose Display Board, Practi-metric Gummed Paper Shapes, Visi-Clear Fractions Rubber Stamps, Visual Fractions Apparatus*

Taskmaster Ltd, *Fraction and Geometric Pieces, Pinboards*

Chapter 15

Graphs

For the teacher

Children will have made and used different types of graphs since they started school. In this chapter the emphasis is on the accuracy of drawing and obtaining information from graphs. The children are encouraged to carry out their own investigations and to illustrate the collected data with graphs. Using straight lines as conversion graphs highlights the need to interpolate or read between marked points. The chapter ends with an illustration of a different kind of graph – a nomogram.

Summary of the stages

1 Using a graph
2 Illustrating an investigation
3 Temperature graphs
4 Conversion graphs
5 A nomogram

Vocabulary

Popular, receipts, investigate, investigation, horizontal, vertical, column, maximum, conversion, currency, exchange, nomogram, axis, axes.

Equipment and apparatus

Centimetre-squared paper, rulers, crayons.

Working with the children

1 Using a graph

Graphs are not drawn in order to produce something to put up on the wall. They have a purpose; they are very much a part of the world of communication. Most newspapers contain at least one graphical illustration which is there, not so that we can admire the skill of its composer, but to show, in a simple way, complex numerical data. There are two essential components of every graph. *Firstly* it should be clearly titled; a graph without a title is as useless as an envelope without an address. *Secondly* its axes should be labelled and given appropriate scales. In the classroom, however, graphs should always have a third component – a piece of writing describing the graph, a set of questions to be answered by using it, or both.

A graph showing attendances at the school play has been drawn for the children to use to answer questions. These range from the straightforward reading of column heights to questions involving fractions and computation. The pupils are then asked to draw a graph from given data and to make up their own questions.

Examples of graphs from newspapers and magazines should be collected and used to extract information.

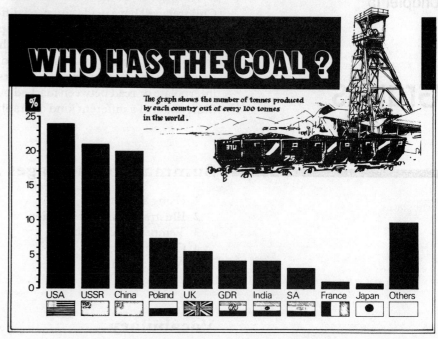

2 Illustrating an investigation

The collection and classification of data is an important mathematical skill which has uses in other subjects (e.g. science, geography). In the example given the children are reminded how to tally, putting marks into sets of five by drawing the fifth across the other four – ̶I̶I̶I̶I̶. Counting in fives is fairly easy so the number of items recorded is quickly calculated. The pupils are then taken step by step through the drawing of a block graph illustrating the information. On completion of the drawing they are asked to use the graph by writing about it.

The final question of this page is open-ended, asking the pupils to choose other topics, collect information and draw a graph of their results. (*Nuffield Maths 5 Spiritmasters*, Grid 13.) Some possible topics are listed below –

Names of 'Pop' groups	Popularity in class
Names of countries	Populations
Names of countries	Land areas
Names of towns and cities	Populations
Names of mountains	Heights
Speed of cars in m.p.h.	Stopping distance
Makes of cars	Number of parents owning each
Makes of cars	Prices
Makes of cars	Approximate number of miles per gallon
Football teams	Goals for and against
Days of the month	Maximum and minimum outside temperatures
Days of the month	Rainfall
Christian names	Number in class each name

reasons why a real unit is not used, firstly, in the fluctuating money markets of today rates of exchange vary so quickly that a question would soon be out of date and secondly, in real units awkward numbers occur (e.g. £1 . . . 10.76 francs).

The pupil is asked to look closely at the axes to appreciate the scale used to represent units before using the graph to convert from one currency to the other.

The second example has a graph for the children to draw and questions to answer on it. Here, a year later, the rate of exchange has altered from £1 = 8 mu to £1 = 10 mu. This is a good class discussion point – does this change benefit us or does it benefit the Upper Mathematicans?

Examples are set of exchange outside the range of the graph. These can be done by partitioning the amount to be exchanged (e.g. £19 = £5 + £5 + £5 + £4).

5   A nomogram

This is included as the last exercise in the *Pupils' Book* to show them that all graphs are not based on two axes. The nomogram drawn is a simple one and may encourage the more able to attempt to make others. (*Nuffield Maths 5 Spiritmasters*, Grid 18.)

Appendix 1

Types of graphical representation

Although the children have been drawing graphs for some time it is only now that we begin to use them in detail. Throughout the chapter the block graph has been used but many other types of representation are available and should be encouraged.

1   Block charts

Each object is represented by a 'block', a square (or rectangle) on the paper which is lined round and then shaded in.

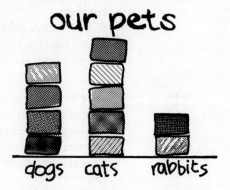

2   Block graphs or bar charts

Instead of drawing separate symbols, long rectangles (bars) of the same width can represent the data. It is the length (height) of the column which

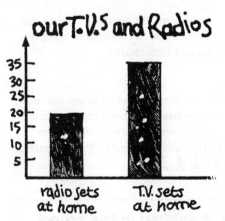

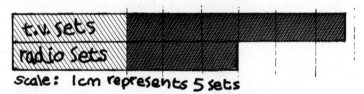

'catches the eye' and gives the information required. Sometimes the columns are separated, the points between the bars being meaningless, or the bars may be adjacent. The vertical axis may detail the numbers to scale, the details being shown below the illustration.

The height or length tells you on sight that the number of T.V. sets is almost double the number of radio sets.

3 Bar line charts

This type of graph is made up of lines instead of rectangles and the lines may be arranged horizontally or vertically. The numbers concerned are represented by *length*.

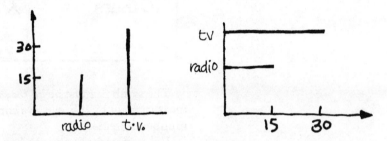

Bar lines are easy to use when approximating and 'rounding off' numbers where decimal parts of a centimetre are used. For example, a line to represent 387 people could be drawn 3.9 cm long (390 people, 100 people represented by one cm) and 952 people would be represented by a line 9.5 cm long.

4 Pie charts (representing data with a circle)

These graphs use the idea of sector representation, and are usually employed to show the proportions in which a whole is made up from its parts. If there are not too many sub-divisions and the sub-divisions are not too small, the child can observe the ratio between the different parts or what fraction each part represents of the whole. *All* sub-divisions of the subject under consideration must be included in the data. For example, for the total household expenditure, *all* sub-divisions of expenditure must be included.

This type of representation is not recommended for wide use, but at a later stage it can bring in a number of ideas:

a   angular measurement and use of protractor.

b   proportion. Area of sector is proportional to the size of the angle at the centre; for example, total number represented by 360 degrees is say 900, then

150 is represented by $\frac{150}{900} \times 360$ degrees, i.e. 60 degrees.

c   the size of the circle is unimportant, i.e. the length of the arms of an angle does not condition the size of the angle.

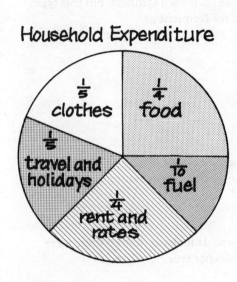

This form of representation has its limitations, the main one being the difficulty in appraising small variations in sector areas compared with the corresponding ease when using rectangles in block charts.

5 Isotypes

An isotype is a picture-diagram. It is often referred to as a pictogram or ideograph, and is used widely in advertisements. A symbol represents a certain number of people or objects. The danger in this representation comes when fractions of a unit are used, for example, each 🧍 represents 10 men so 35 men will be represented by 🧍🧍🧍🧍.

The symbols must be *all the same size and spaced equally* so that the measurement by the 'eye' is a correct one. One symbol can stand for any number of items.

It is important to observe how a picture can represent a completely different quantity, for example:

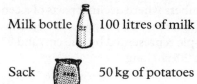

Milk bottle   100 litres of milk

Sack   50 kg of potatoes

With discussion, the children can devise their own symbols, but this type of representation is not recommended for frequent use.

Appendix 2

Common errors in graph drawing

This section deals with some of the things that can go wrong and some of the places where special care is needed.

What's wrong with this?

Clearly the boxes should be the same size. It looks as though there were more boys than girls, but of course this is not true.

What's wrong with this?

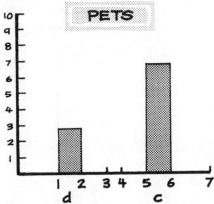

1 There are no titles giving meaning to the horizontal and vertical axes. Does 'd' stand for dogs, ducks, dormice . . . ?
2 An arrow should be shown on the vertical axis following the last number to show that the numbers could be continued if necessary.
3 Writing in numbers on the horizontal axis is incorrect; only dogs and cats are being considered.
4 The bars should be evenly spaced, in spite of the fact that the intervals between them are meaningless.

This would have been a better presentation:

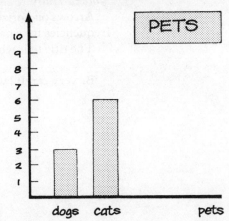

Horizontal axis
Care should be taken when marking this axis.

a Order not required
Any order is used. We could have shown the 'bird' column as the first, then the 'cat' column.

Note carefully:
Arrow used with vertical axis. We could have continued to 20, 25 . . . had it been necessary.
No arrow used with horizontal axis. We are concerned with sets of dogs, birds, cats not numbers.
Equal intervals used on the horizontal axis.
Titles are stated. There is no doubt what our picture is meant to convey to the child.

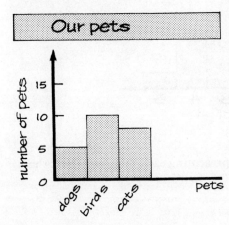

b Order required

The shoe sizes are arranged in order on the horizontal axis. This order assists quick interpretation.

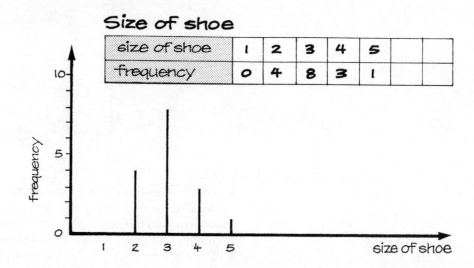

Size of shoe

size of shoe	1	2	3	4	5		
frequency	0	4	8	3	1		

The frequency is the number of times a particular size of shoe occurs in the sample.

Note carefully:

Arrows on horizontal and vertical axes. More shoe sizes and higher frequencies for a particular size could be shown if data turned up.

The title tells what the picture is about.

Be very careful where you label the horizontal axis for a block graph.

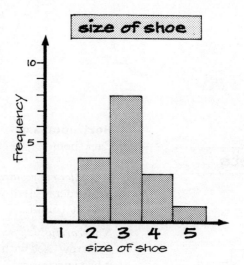

Note the position of the numeral representing each shoe size – near the centre of the rectangle representing the size.

What's wrong with the following?

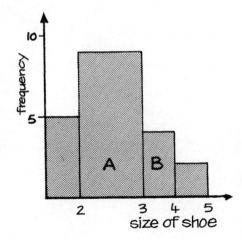

There were no shoes size 1, so that has been ignored, but such negative information can be important.

The intervals are varied. The widths of the rectangles A and B should be equal.

The numeral representing each shoe size is placed incorrectly on the horizontal axis (see the illustration above).

Study this istotype

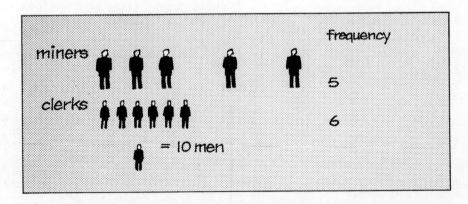

1 The symbols should be the same size, and equally spaced. There are children who would look at this representation and state that there were more miners than clerks.

2 How can ![icon] equal 10 men?

This should have been displayed as:

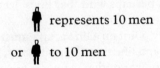

What's wrong with this pie chart?

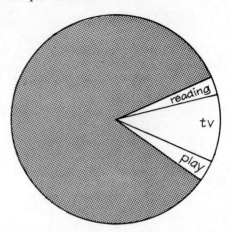

1 There is no explanation of the whole, which should be given in the title or at the side. The biggest sector is unlabelled.
2 Each sector should show a fraction, percentage or a decimal according to how the calculation has been made.
3 The divisions for play and reading are too small for a pie chart representation. It is impossible to see at a glance the ratio between the parts and what approximate fraction each part represents of the total.

(Of course, if the purpose were just to show the inordinate share of time spent in watching T.V., this representation would serve well enough.)

To join or not to join?

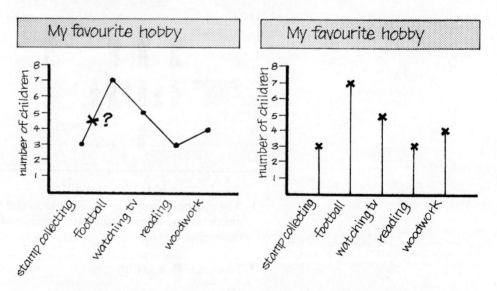

The graph should have consisted simply of the isolated points marked with crosses, perhaps with 'bar lines' leading up to them to emphasise their positions.

Every point on a line graph must mean something, so it is absurd to 'join up' a graph like this. *A point, say, halfway along one of the joining lines has no interpretation whatsoever – unless it means that $4\frac{1}{2}$ boys either collect footballs or kick stamps!

What's wrong with this?

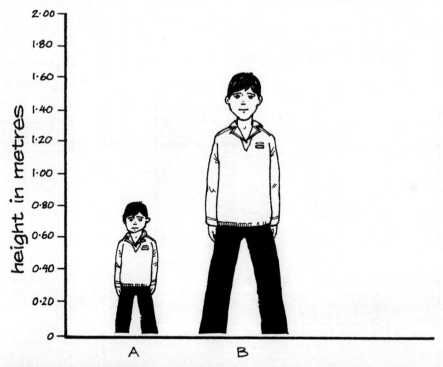

B is twice the height of A and this representation is misleading because the area of B is *four* times the area of A.

When charts are drawn using picture symbols representing area or volume, they must be observed very closely. A bar line chart would have been preferable in this case.

Conclusions

Care must be taken to distinguish between what is *certain* from a graph and what *seems likely*. For example, from one graph it may be possible to say with certainty:

'There are more boys than girls in the class'

but from another graph, in which sales are shown as having gone up over, say, five months, it would at most be possible to predict:

'Sales are *likely* to continue to rise'.

Pages from the Pupils' Book and Spiritmasters

69

Chapter 15: Graphs

1 This graph shows how many people attended the school play each night. The hall will hold 100 people. Tickets were 30p for adults and 20p for children.
   a How many more were at the play on Friday than on Tuesday?
   b Which was the most popular night?
   c Which was the least popular night?
   d What was the total attendance for the five nights?
   e What fraction of the hall was full on Tuesday night?
   f What fraction of the hall was empty on Monday night?
   g If 32 children were at the play on Friday what was the total amount of money taken on Friday?

Graph showing attendance at school play.

2 a This table shows attendances each night. Copy and complete it.

Attendances	Monday	Tuesday	Wednesday	Thursday	Friday
Adults			59	58	
Children	30	28			32

   b What were the total receipts for the week?

3 Use these figures to draw a graph like the one above.

Attendances out of 32 for class 4	Monday		Tuesday		Wednesday		Thursday		Friday	
	a.m.	p.m.	a.m.	p.m.	a.m.	p.m.	a.m.	p.m.	a.m.	p.m.
	26	27	27	27	29	30	31	29	32	32

When you have drawn it make up some problems like those in question 1.

70

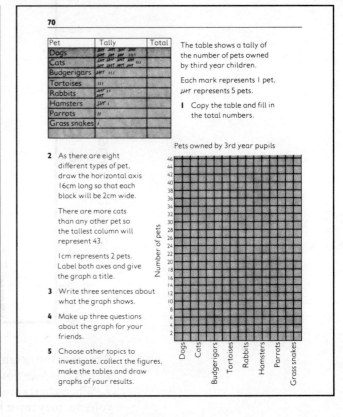

Pet	Tally	Total
Dogs	卌 卌 卌 卌	
Cats	卌 卌 卌 卌 lll	
Budgerigars	卌 lll	
Tortoises	lll	
Rabbits	卌 ll / 卌 ll	
Hamsters	卌 l	
Parrots	ll	
Grass snakes	l	

The table shows a tally of the number of pets owned by third year children.

Each mark represents 1 pet, 卌 represents 5 pets.

1 Copy the table and fill in the total numbers.

Pets owned by 3rd year pupils

2 As there are eight different types of pet, draw the horizontal axis 16cm long so that each block will be 2cm wide.

There are more cats than any other pet so the tallest column will represent 43.

1cm represents 2 pets. Label both axes and give the graph a title.

3 Write three sentences about what the graph shows.

4 Make up three questions about the graph for your friends.

5 Choose other topics to investigate, collect the figures, make the tables and draw graphs of your results.

71

Temperature graphs

The following table gives the average maximum temperatures in °C for London.

Jan	Feb	Mar	Apl	May	June	July	Aug	Sept	Oct	Nov	Dec
6	7	10	13	17	20	22	21	19	14	10	7

These can be shown on a graph.

Average maximum temperatures in London

1 Which month had the highest temperature and which the lowest?
2 The difference between the highest and lowest temperatures is called the **range**. What is the range of temperatures in London?
3 Which months have the same temperatures?

A travel magazine gives the average maximum temperatures in °C for 3 Greek islands as:

	Jan	Feb	Mar	Apl	May	June	July	Aug	Sept	Oct	Nov	Dec
Limnos	11	12	13	18	23	27	30	30	26	22	16	13
Naxos	15	15	16	20	23	26	27	28	26	24	20	17
Zakynthos	14	14	16	22	32	29	32	32	29	25	20	16

4 Using a scale of 2cm for each month on the horizontal axis and 1cm for 2°C on the vertical axis, draw a separate graph for each island.
5 What is the range of temperatures for each island?
6 Which months in London are colder than the lowest month in Naxos?
7 How much warmer than London is Zakynthos in August?

72

Conversion graphs

In Britain the unit of currency is the pound (£). Other countries have different units of currency. In Upper Mathematica the unit of currency is the mult (mu). A visitor from Britain, last year, received 8mu for each £1.

The graph shows this conversion. The pounds have been marked on the horizontal axis and the mults on the vertical axis.

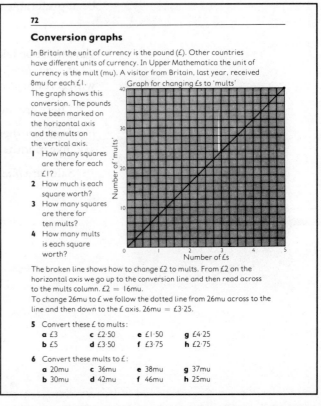

Graph for changing £s to 'mults'

1 How many squares are there for each £1?
2 How much is each square worth?
3 How many squares are there for ten mults?
4 How many mults is each square worth?

The broken line shows how to change £2 to mults. From £2 on the horizontal axis we go up to the conversion line and then read across to the mults column. £2 = 16mu.
To change 26mu to £ we follow the dotted line from 26mu across to the line and then down to the £ axis. 26mu = £3·25.

5 Convert these £ to mults:
   a £3   c £2·50   e £1·50   g £4·25
   b £5   d £3·50   f £3·75   h £2·75

6 Convert these mults to £:
   a 20mu   c 36mu   e 38mu   g 37mu
   b 30mu   d 42mu   f 46mu   h 25mu

73

Visitors to Upper Mathematica this year found that they received 10 mults for £1.

I Copy and complete:

a £1 = 10mu   **c** £3 = ▨   **e** £5 = ▨
b £2 = ▨   **d** £4 = ▨   **f** £0 = ▨

2 Draw axes as in the previous conversion graph but make the vertical axis taller, taking it to 50 mults.

3 Plot the points in question 1 and join them with a straight line.

4 Using your graph, copy and complete this table:

£	I		4		I·50		2·75		4·75	
mults		30		50		35		32½		37½

To exchange larger amounts than £5 we can either say:
£11 = £5 + £5 + £1 = 50mu + 50mu + 10mu = 110mu
or £18 = 6 × £3   = 6 × 30mu   = 180mu

5 Using your graph copy, complete this table:

£	7	9	13	15	24	30
Mults						

6 Last year I bought a map in Upper Mathematica for 12mu. How many £ was that? (Use last year's graph!)

7 This year it is still 12mu. What would I have saved in £ if I had bought it this year?

8 A meal cost me 32mu last year. If the price is still the same this year what do I save in £ if I have a similar meal?

9 When I went to Upper Mathematica last year my holiday cost me 1 856 mults. How many pounds was equal to that amount?
If my holiday this year will cost the same number of mults how many pounds will it cost?
What amount (in pounds) do I save this year?

74

A different sort of graph—a nomograph

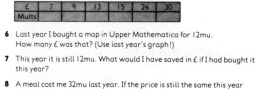

A   B   C

This graph is designed to help us to add or subtract.

I On centimetre squared paper draw three vertical lines each 20cm long and 3cm apart. Letter them **A**, **B** and **C**.

2 Number them as in the diagram with 2cm for each unit on the **A** and **C** columns and 1cm for each unit on the **B** column.

3 To add 5 and 4 place a ruler across the chart (see diagram) from 5 in column **A** to 4 in column **C**. The answer 9 is in the middle column.

4 Use the nomogram to complete:

a 1 + 2 =   **f** 4½ + 3½ =
b 3 + 5 =   **g** 7 + 2½ =
c 7 + 2 =   **h** 3½ + 5 =
d 3½ + 2½ =   **i** 6½ + 1 =
e 6½ + ½ =   **j** 5½ + 2½ =

5 Look at the broken line. Can you see how we could read 9 − 5 = 4? Which column is the answer in?

6 Complete using the nomogram:

a 7 − 5 =   **e** 8½ − 3½ =
b 8 − 2 =   **f** 6½ − 5½ =
c 6 − 4 =   **g** 8 − 3½ =
d 7½ − 2½ =   **h** 6½ − 4 =

References and resources

Nuffield Mathematics Teaching Project, *Pictorial Representation* [1] Chambers/Murray 1967 (See Introduction page xi.)

Williams E. M. and Shuard, H. *Primary Mathematics Today*, Third Edition (Chapter 28), Longman Group Ltd 1982

Practice 15
Chapter 15: Graphs

Name _____

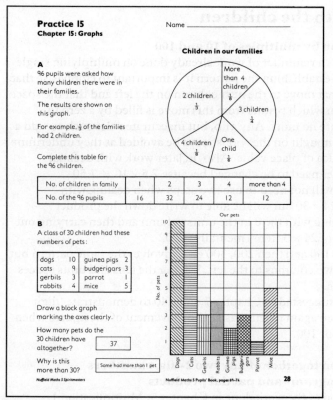

Children in our families

A
96 pupils were asked how many children there were in their families.

The results are shown on this graph.

For example, ⅓ of the families had 2 children.

Complete this table for the 96 children.

No. of children in family	I	2	3	4	more than 4
No. of the 96 pupils	16	32	24	12	12

B
A class of 30 children had these numbers of pets:

dogs	10	guinea pigs	2
cats	9	budgerigars	3
gerbils	3	parrot	1
rabbits	4	mice	5

Draw a block graph marking the axes clearly.

How many pets do the 30 children have altogether?   37

Why is this more than 30?   Some had more than 1 pet

Nuffield Maths 5 Spiritmasters

Nuffield Maths 5 Pupils' Book, pages 69–74   **28**

Chapter 16

Multiplication 2

For the teacher

After looking at multiplication by 10, 100 and their multiples, the work in Chapter 5, Multiplication 1, is extended to the multiplication together of two-digit numbers using diagrams and the traditional long multiplication layout.

Summary of the stages

1 Multiplication by multiples of 10 and 100
2 Multiplication together of any two 2-digit numbers using area diagrams and partial products
3 Word and pattern problems

Vocabulary

Column, zero, multiples, digits, partial products.

Equipment and apparatus

Graph paper (small squares), 2-centimetre-squared paper.

Working with the children

1 Multiplication by multiples of 10 and 100

The first exercise is a reminder of work already done on multiplying single digits by 10. In re-establishing the pattern it is important to emphasise that it is really *figures* that move to the next column on the left and that the space in the units column which results from this move is filled by a zero to signify that there are no units. Any glib, but inaccurate rules such as 'add a nought' or 'stick a nought on the end' should be avoided as they undermine the fundamental idea of place value. Also, in later work with decimal numbers, such rules need to be changed because $7.5 \times 10 \neq 7.50$.

Many children will not need to use two steps when dealing with examples such as 24×30 but will be able to write down the product straight away starting with the 0 in the units column and then carrying out the multiplication of 24×3 (tens) mentally: 720.

Multiplying by 100 and then 200, 300 etc., involves the same pattern but this time moving two columns to the left, leaving the tens and units spaces to be filled by zeros.

The apparatus suggested in Chapter 7, page 40 to demonstrate 'digit shift' should be used again to emphasise the movement of the figures when multiplying by 10 or 100.

2 Multiplication together of any two 2-digit numbers using area diagrams and partial products

This is an extension of the work done in Chapter 5, Multiplication 1.

Again, if the children are to draw an accurate diagram, they need graph paper with small squares, for example, 38 × 27:

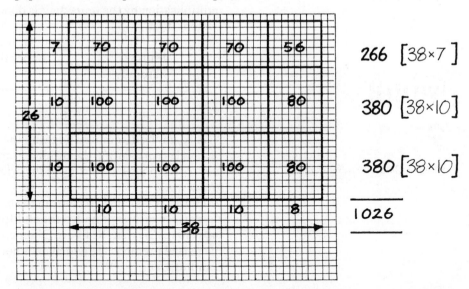

The drawing of an accurate diagram like this is not really necessary and a free-hand sketch is all that is needed, for example 34 × 22:

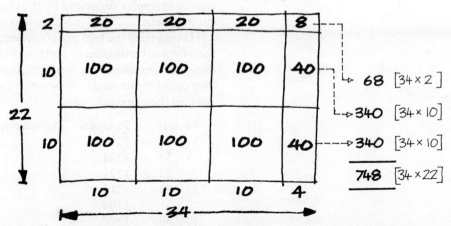

If children find the drawing of a sketch difficult, they can either continue to use small square graph paper or 2-centimetre-squared paper with each square representing 10 × 10.

The plan is then refined to reduce the number of partial products, leading to the traditional layout. For example, 37 × 26:

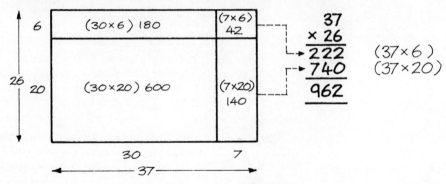

Multiplying by a two-digit number is seen as an extension of the previous work on multiplication by a single digit. For this reason, multiplication by the units digit is done first, followed by multiplication by the tens digit.

$$
\begin{array}{r} 34 \\ \times\ 7 \\ \hline 238 \\ {\scriptstyle 2} \end{array}
\quad \text{leads on naturally to} \quad
\begin{array}{r} 34 \\ \times 27 \\ \hline 238 \\ 680 \\ \hline 918 \end{array}
\quad \text{rather than} \quad
\begin{array}{r} 34 \\ \times 27 \\ \hline 680 \\ 238 \\ \hline 918 \end{array}
$$

Also, in the diagrammatic form, the measurements of the sides of the rectangle representing the product start from the bottom left-hand corner. This corresponds to starting at the origin when measuring along the axes in graph work.

However, as can be seen from the example above, either method leads to the correct answer provided the children are familiar enough with place value to realise that the figure '2' represents '2 tens'. The order in which the partial products are added does not matter because addition is commutative.

3   Word and pattern problems

The problems presented in the *Pupils' Book* should be supplemented by similar examples suggested by the children or taken from their environment.

Having tried the two pattern questions, some children may wish to start their own investigation of patterns in multiplication. Alternatively, once they have practised long multiplication for the first few cases of a pattern, they could try to predict how it will continue, perhaps using a calculator to confirm their predictions.

Factors	Products	Differences
35×35 =	1225	
36×34 =	1224	1
37×33 =	1221	3
38×32 =	1216	5
39×31 =	1209	7
40×30 =	1200	9

Pages from the Pupils' Book and Spiritmasters

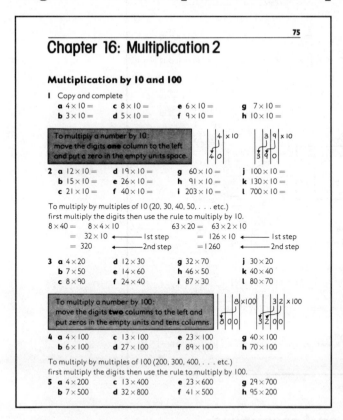

75

Chapter 16: Multiplication 2

Multiplication by 10 and 100

1 Copy and complete
a $4 \times 10 =$ **c** $8 \times 10 =$ **e** $6 \times 10 =$ **g** $7 \times 10 =$
b $3 \times 10 =$ **d** $5 \times 10 =$ **f** $9 \times 10 =$ **h** $10 \times 10 =$

> To multiply a number by 10:
> move the digits **one** column to the left
> and put a zero in the empty units space.

2 a $12 \times 10 =$ **d** $19 \times 10 =$ **g** $60 \times 10 =$ **j** $100 \times 10 =$
b $15 \times 10 =$ **e** $26 \times 10 =$ **h** $91 \times 10 =$ **k** $130 \times 10 =$
c $21 \times 10 =$ **f** $40 \times 10 =$ **i** $203 \times 10 =$ **l** $700 \times 10 =$

To multiply by multiples of 10 (20, 30, 40, 50, . . . etc.)
first multiply the digits then use the rule to multiply by 10.

$8 \times 40 = \quad 8 \times 4 \times 10 \qquad\qquad 63 \times 20 = \quad 63 \times 2 \times 10$
$\quad = \quad 32 \times 10 \leftarrow \text{1st step} \qquad = \quad 126 \times 10 \leftarrow \text{1st step}$
$\quad = \quad 320 \leftarrow \text{2nd step} \qquad\quad = \quad 1260 \leftarrow \text{2nd step}$

3 a 4×20 **d** 12×30 **g** 32×70 **j** 30×20
b 7×50 **e** 14×60 **h** 46×50 **k** 40×40
c 8×90 **f** 24×40 **i** 87×30 **l** 80×70

> To multiply a number by 100:
> move the digits **two** columns to the left and
> put zeros in the empty units and tens columns.

4 a 4×100 **c** 13×100 **e** 23×100 **g** 40×100
b 6×100 **d** 27×100 **f** 89×100 **h** 70×100

To multiply by multiples of 100 (200, 300, 400, . . . etc.)
first multiply the digits then use the rule to multiply by 100.

5 a 4×200 **c** 13×400 **e** 23×600 **g** 29×700
b 7×500 **d** 32×800 **f** 41×500 **h** 95×200

76

Here is a diagram which shows 16×14

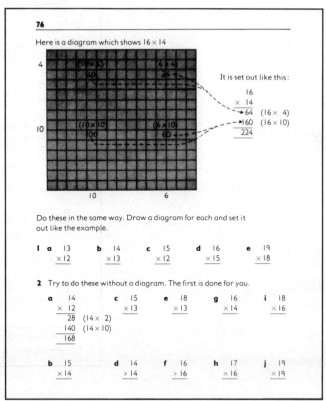

It is set out like this:

$$\begin{array}{r} 16 \\ \times\ 14 \\ \hline 64 \quad (16 \times 4) \\ 160 \quad (16 \times 10) \\ \hline 224 \end{array}$$

Do these in the same way. Draw a diagram for each and set it
out like the example.

1 a $\begin{array}{r} 13 \\ \times 12 \end{array}$ **b** $\begin{array}{r} 14 \\ \times 13 \end{array}$ **c** $\begin{array}{r} 15 \\ \times 12 \end{array}$ **d** $\begin{array}{r} 16 \\ \times 15 \end{array}$ **e** $\begin{array}{r} 19 \\ \times 18 \end{array}$

2 Try to do these without a diagram. The first is done for you.

a $\begin{array}{r} 14 \\ \times 12 \\ \hline 28 \quad (14 \times 2) \\ 140 \quad (14 \times 10) \\ \hline 168 \end{array}$ **c** $\begin{array}{r} 15 \\ \times 13 \end{array}$ **e** $\begin{array}{r} 18 \\ \times 13 \end{array}$ **g** $\begin{array}{r} 16 \\ \times 14 \end{array}$ **i** $\begin{array}{r} 18 \\ \times 16 \end{array}$

b $\begin{array}{r} 15 \\ \times 14 \end{array}$ **d** $\begin{array}{r} 14 \\ \times 14 \end{array}$ **f** $\begin{array}{r} 16 \\ \times 16 \end{array}$ **h** $\begin{array}{r} 17 \\ \times 16 \end{array}$ **j** $\begin{array}{r} 19 \\ \times 19 \end{array}$

77

This diagram shows 38×27

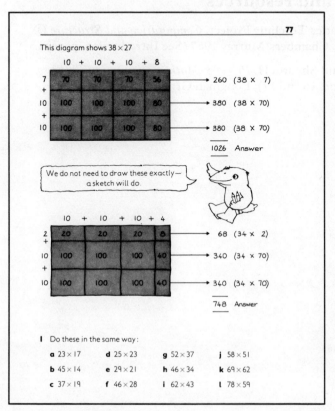

> We do not need to draw these exactly —
> a sketch will do.

1 Do these in the same way:

a 23×17 **d** 25×23 **g** 52×37 **j** 58×51
b 45×14 **e** 29×21 **h** 46×34 **k** 69×62
c 37×19 **f** 46×28 **i** 62×43 **l** 78×59

78

Here is a plan divided into fewer parts. It shows 37×26.

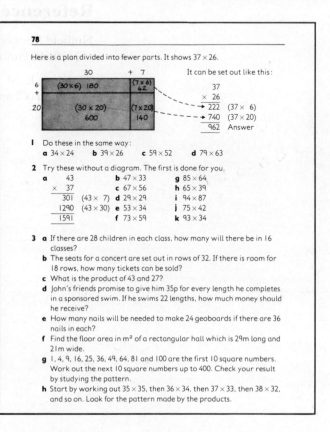

It can be set out like this:

$$\begin{array}{r} 37 \\ \times\ 26 \\ \hline 222 \quad (37 \times 6) \\ 740 \quad (37 \times 20) \\ \hline 962 \quad \text{Answer} \end{array}$$

1 Do these in the same way:
a 34×24 **b** 39×26 **c** 59×52 **d** 79×63

2 Try these without a diagram. The first is done for you.

a $\begin{array}{r} 43 \\ \times\ 37 \\ \hline 301 \quad (43 \times 7) \\ 1290 \quad (43 \times 30) \\ \hline 1591 \end{array}$ **b** 47×33 **g** 85×64
c 67×56 **h** 65×39
d 29×29 **i** 94×87
e 53×34 **j** 75×42
f 73×59 **k** 93×34

3 a If there are 28 children in each class, how many will there be in 16 classes?
b The seats for a concert are set out in rows of 32. If there is room for 18 rows, how many tickets can be sold?
c What is the product of 43 and 27?
d John's friends promise to give him 35p for every length he completes in a sponsored swim. If he swims 22 lengths, how much money should he receive?
e How many nails will be needed to make 24 geoboards if there are 36 nails in each?
f Find the floor area in m² of a rectangular hall which is 29m long and 21m wide.
g 1, 4, 9, 16, 25, 36, 49, 64, 81 and 100 are the first 10 square numbers. Work out the next 10 square numbers up to 400. Check your result by studying the pattern.
h Start by working out 35×35, then 36×34, then 37×33, then 38×32, and so on. Look for the pattern made by the products.

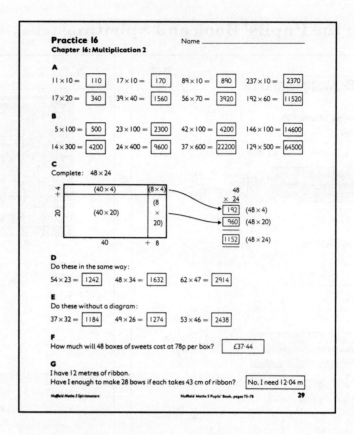

Practice 16
Name _____

Chapter 16: Multiplication 2

A

$11 \times 10 =$ [110] $17 \times 10 =$ [170] $89 \times 10 =$ [890] $237 \times 10 =$ [2370]

$17 \times 20 =$ [340] $39 \times 40 =$ [1560] $56 \times 70 =$ [3920] $192 \times 60 =$ [11520]

B

$5 \times 100 =$ [500] $23 \times 100 =$ [2300] $42 \times 100 =$ [4200] $146 \times 100 =$ [14600]

$14 \times 300 =$ [4200] $24 \times 400 =$ [9600] $37 \times 600 =$ [22200] $129 \times 500 =$ [64500]

C

Complete: 48×24

	(40×4)	(8×4)
	(40×20)	(8×20)

$40+8$

$\begin{array}{r} 48 \\ \times\ 24 \\ \hline \end{array}$
[192] (48×4)
[960] (48×20)
[1152] (48×24)

D

Do these in the same way:

$54 \times 23 =$ [1242] $48 \times 34 =$ [1632] $62 \times 47 =$ [2914]

E

Do these without a diagram:

$37 \times 32 =$ [1184] $49 \times 26 =$ [1274] $53 \times 46 =$ [2438]

F

How much will 48 boxes of sweets cost at 78p per box?   [£37·44]

G

I have 12 metres of ribbon.
Have I enough to make 28 bows if each takes 43 cm of ribbon?   [No, I need 12·04 m]

Nuffield Maths 5 Spiritmasters   *Nuffield Maths 5 Pupils' Book, pages 75-78*   **29**

References and resources

Nuffield Mathematics Teaching Project, *Computation and Structure* ③
   Nuffield Guide, Chambers/Murray 1967 (See Introduction, page xi.)

Williams, E. M. and Shuard, H. *Primary Mathematics Today* Third
   Edition (Chapters 16 and 17), Longman Group Ltd 1982

Chapter 17

Shape 2

For the teacher

The format of this chapter follows that of Chapter 2, Shape 1, but deals with the angles and properties of quadrilaterals. The final, practical section on simple tessellations provides reinforcement of work done on angles.

Summary of the stages

1 Angles of a quadrilateral
2 Some special quadrilaterals
3 Tessellation

Vocabulary

Quadrilateral, square, rectangle, rhombus, parallelogram, tessellation, right-angled, equilateral, scalene, bisect.

Equipment and apparatus

Stiff paper or thin card, centimetre-squared paper, coloured sticky paper or plastic shapes, scissors.

Working with the children

1   Angles of a quadrilateral

The 'tearing and fitting' gives a practical demonstration of the angle-sum of a quadrilateral. Tearing, rather than cutting off the corners is suggested deliberately since a straight cut could lead to confusion when fitting the corners together to form 360° around a point. Clear marking of the angles before tearing is also helpful.

Some children will be able to appreciate that by drawing one diagonal of the quadrilateral, two triangles are formed. The sum of the angles of the quadrilateral can now be seen as the same as the sum of the angles of the two triangles:

$$a+b+c = \underline{180°}$$
$$x+y+z = \underline{180°}$$

Sum of angles of
quadrilateral $= 360°$

As in the case of triangles, finding the missing angle of a quadrilateral is similar to a 'missing number' problem:

$$95° + 80° + 115° + \boxed{}° = 360°$$

The missing angle is $70°$.

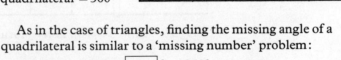

95	360
+80	−290
115	70
290	

2 Some special quadrilaterals

In this section special attention will need to be given to the correct meaning (and spelling!) of such words as square, rhombus, rectangle, parallelogram, diagonal, symmetry. All the examples presented in the *Pupils' Book* are in fact parallelograms – they all have opposite sides equal and parallel. But in this context the word parallelogram is used for the *general case*, that is a parallelogram which does not necessarily have four equal sides (in which case it is a rhombus) or four right angles (in which case it is a rectangle). Similarly, a square can be seen either as a special rhombus or a special rectangle. A set diagram helps to make these points clearer:

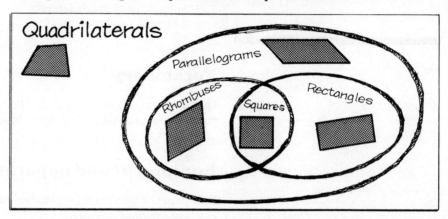

As an alternative to the table in the *Pupils' Book*, the following chart could be discussed and completed with a group of children, after pointing out that *bisect* means to cut into two equal parts:

	opposite sides equal?	opposite sides parallel?	opposite angles equal?	diagonals bisect each other?	diagonals at right angles?	all sides equal?	one angle a right angle?	diagonals equal?
parallelogram	Yes	Yes	Yes	Yes	not always (if so it's a rhombus)	not always (if so it's a rhombus)	not always (if so it's a rectangle)	not always (if so it's a rectangle)
rectangle	Yes	Yes	Yes	Yes	not always (if so it's a square)	not always (if so it's a square)	Yes	Yes
rhombus	Yes	Yes	Yes	Yes	Yes	Yes	not always (if so it's a square)	not always (if so it's a square)
square	Yes	Yes	Yes	Yes	Yes	Yes	Yes	Yes

(*Nuffield Maths 5 Spiritmasters*, Grids 16 and 17.)

The exercises involving drawing shapes, marking right angles, pairs of equal angles, and calculating missing angles reinforce the work done so far.

3 Tessellation

After looking for examples of tessellating shapes in the environment, the children are encouraged to investigate patterns of their own – in the first instance using a set of 'tiles', all the same shape and size. With reasonable care and accuracy they should be able to conclude that all triangles tessellate and all quadrilaterals tessellate. The patterns made will also reinforce the angle-sum of triangles and quadrilaterals:

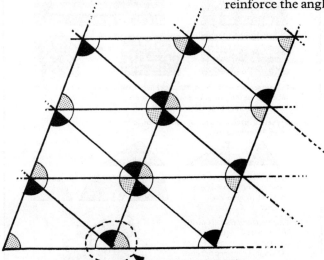

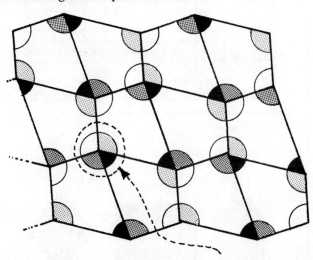

The 3 angles of a triangle together make 180°.

The 4 angles of a quadrilateral together make 360°.

Pages from the Pupils' Book and Spiritmasters

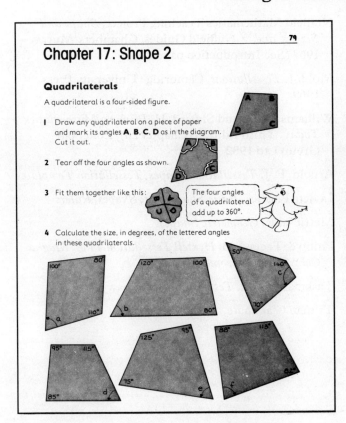

79

Chapter 17: Shape 2

Quadrilaterals

A quadrilateral is a four-sided figure.

1 Draw any quadrilateral on a piece of paper and mark its angles **A**, **B**, **C**, **D** as in the diagram. Cut it out.

2 Tear off the four angles as shown.

3 Fit them together like this:

The four angles of a quadrilateral add up to 360°.

4 Calculate the size, in degrees, of the lettered angles in these quadrilaterals.

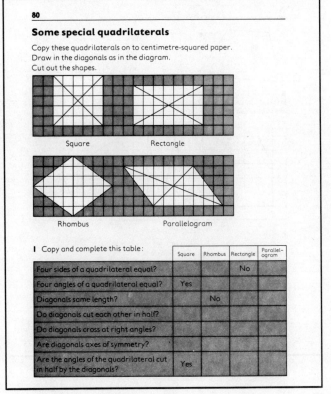

80

Some special quadrilaterals

Copy these quadrilaterals on to centimetre-squared paper.
Draw in the diagonals as in the diagram.
Cut out the shapes.

Square   Rectangle

Rhombus   Parallelogram

1 Copy and complete this table:

	Square	Rhombus	Rectangle	Parallel-ogram
Four sides of a quadrilateral equal?			No	
Four angles of a quadrilateral equal?	Yes			
Diagonals same length?		No		
Do diagonals cut each other in half?				
Do diagonals cross at right angles?				
Are diagonals axes of symmetry?				
Are the angles of the quadrilateral cut in half by the diagonals?	Yes			

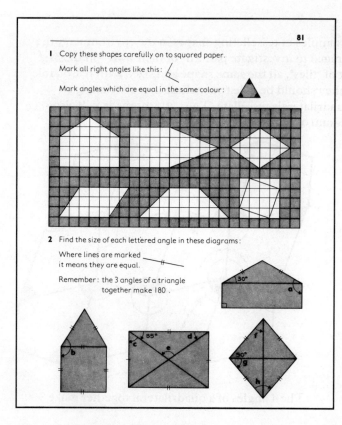

81

1 Copy these shapes carefully on to squared paper.

Mark all right angles like this:

Mark angles which are equal in the same colour:

2 Find the size of each lettered angle in these diagrams:

Where lines are marked it means they are equal.

Remember: the 3 angles of a triangle together make 180.

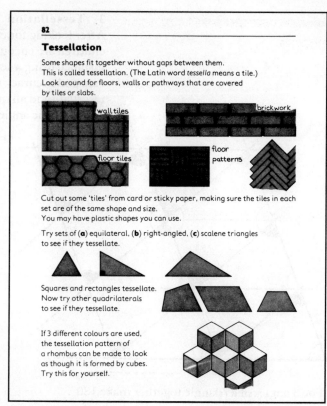

82

Tessellation

Some shapes fit together without gaps between them. This is called tessellation. (The Latin word *tessella* means a tile.)

Look around for floors, walls or pathways that are covered by tiles or slabs.

wall tiles

brickwork

floor tiles

floor patterns

Cut out some 'tiles' from card or sticky paper, making sure the tiles in each set are of the same shape and size.

You may have plastic shapes you can use.

Try sets of (**a**) equilateral, (**b**) right-angled, (**c**) scalene triangles to see if they tessellate.

Squares and rectangles tessellate. Now try other quadrilaterals to see if they tessellate.

If 3 different colours are used, the tessellation pattern of a rhombus can be made to look as though it is formed by cubes. Try this for yourself.

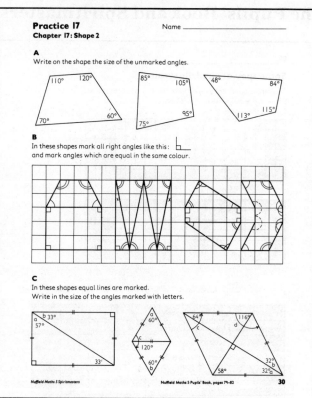

Practice 17
Chapter 17: Shape 2

Name _____

A

Write on the shape the size of the unmarked angles.

B

In these shapes mark all right angles like this: and mark angles which are equal in the same colour.

C

In these shapes equal lines are marked.
Write in the size of the angles marked with letters.

Nuffield Maths 5 Spiritmasters

Nuffield Maths 5 Pupils' Book, pages 79–82

30

References and resources

Nuffield Mathematics Teaching Project, *Shape and Size ▽ and ▽* Nuffield Guides, Chambers/Murray 1967 (See Introduction page xi.)

Mold, J. *Tessellations*, Cambridge University Press 1969

Williams, E. M. and Shuard, H. *Primary Mathematics Today*, Third Edition (Chapter 20), Longman Group Ltd 1982

Arnold, E. J. *Tessellation Shapes, Tessellation Templates*

Invicta Plastics, *Basic Shapes Set, Shapes, Rulers*

Metric-aids, *Geometric Shapes*

Philip & Tacey Ltd, *Hextell Tessellation Tiles, Magna-Cel Magnetic Geometric Shapes*

Taskmaster Ltd, *Tessellations* (H. Shaw)

Triman Classmate, *Tactile Tessellations*

Chapter 18

Division 2

For the teacher

The method of division used in Division 1 (Chapter 7) is extended. Firstly, after revision of multiplication by 100 and multiples of 100, division is carried out by a single digit producing answers greater than 100. Again step by step repeated subtraction is used as this enables the pupils to move on to the next stage more easily.

Secondly, the pupils divide three digits by two. Although this has traditionally been called 'long division' there should be no difference in layout or thinking from the previous examples and, therefore, no special emphasis need be placed on this stage.

Summary of the stages

1 Multiplication by 100 and multiples of 100 leading to division with answers greater than 100
2 Division of three digits by two with divisors less than 20

Vocabulary

Multiples, divide, column, digit, divisor, dividend, quotient, remainder.

Equipment and apparatus

Squared paper, multiplication squares.

Working with the children

1 Multiplication by 100 and multiples of 100 leading to division with answers greater than 100

The various types of apparatus suggested in Division 1 (Chapter 7) can be extended for use when multiplying by 100. Again it must be emphasised that multiplying by 100 is **not** adding two noughts. If the pupil is to understand place value he must see and learn that digits are moved two places to the left and the spaces filled with zeros.
Multiplication by multiples of 100 is done in two steps:

$$38 \times 400 = 38 \times 4 \times 100$$

Multiply the first two numbers $\qquad = 152 \times 100$
Then by the hundred $\qquad = 15\,200$

In the first set of division in this chapter, answers are less than two hundred so multiples of 100 are not used. The second set gives answers greater than 200 but the divisions are necessarily small. Although the pupils may find these comparatively easy, it is the *layout* which is being learned.

2 Division of three digits by two with divisors less than 20

This layout is the same as that used throughout the work done by the pupil on division. Each digit is given its proper name for the place it is in – the '2' is 200 and not as traditional division would have it, '17 into 2 won't go – 17 into 21 . . .', – the '2' changing its value each time but never being the 200 it

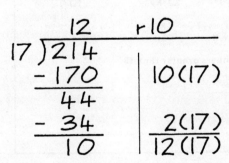

actually is! Another traditional piece of jargon which is not necessary with this method is the 'bringing down of the next figure' (not even in the lift!). When the pupil subtracts 10 (17) then all that remains appears.

As in Division 1 (Chapter 7), whilst the pupil should be encouraged to subtract the highest multiple of the divisor possible, some pupils will do these in smaller steps. It is better that the child produces a correct answer by a long method than that he cannot do the division.

$$
\begin{array}{r}
27 \quad r\,7 \\
16)\overline{439} \\
-160 \quad | \; 10\,(16) \\
\overline{279} \\
-160 \quad | \; 10\,(16) \\
\overline{119} \\
-\;\;64 \quad | \; 4\,(16) \\
\overline{55} \\
-\;\;48 \quad | \; 3\,(16) \\
\overline{7} \quad | \; 27\,(16)
\end{array}
\qquad
\begin{array}{r}
27 \quad r\,7 \\
16)\overline{439} \\
-320 \quad | \; 20\,(16) \\
\overline{119} \\
-112 \quad | \; 7\,(16) \\
\overline{7} \quad | \; 27\,(16)
\end{array}
$$

The chapter ends with a few word problems. It is important that the pupils are encouraged to devise their own 'stories'. For example: 'Make up a story for 376 ÷ 14'.

Pages from the Pupils' Book and Spiritmasters

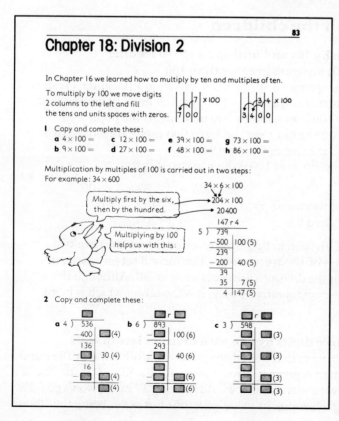

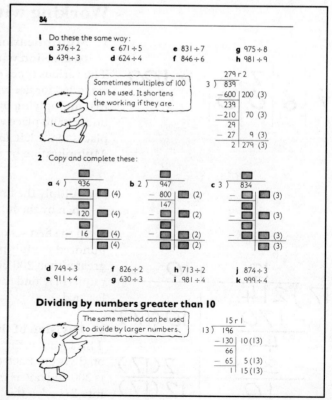

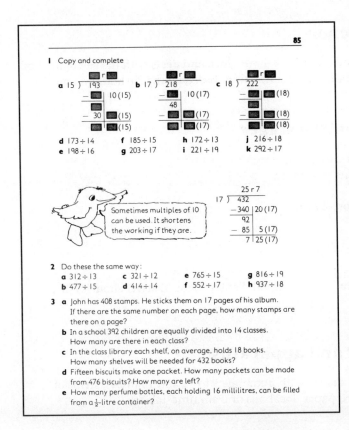

85

1 Copy and complete

a 15) 193

b 17) 218

c 18) 222

d 173 ÷ 14   **f** 185 ÷ 15   **h** 172 ÷ 13   **j** 216 ÷ 18
e 198 ÷ 16   **g** 203 ÷ 17   **i** 221 ÷ 19   **k** 292 ÷ 17

Sometimes multiples of 10 can be used. It shortens the working if they are.

```
        25 r 7
17 ) 432
    − 340  20 (17)
       92
     − 85  5 (17)
        7  25 (17)
```

2 Do these the same way:
a 312 ÷ 13   **c** 321 ÷ 12   **e** 765 ÷ 15   **g** 816 ÷ 19
b 477 ÷ 15   **d** 414 ÷ 14   **f** 552 ÷ 17   **h** 937 ÷ 18

3 **a** John has 408 stamps. He sticks them on 17 pages of his album. If there are the same number on each page, how many stamps are there on a page?
b In a school 392 children are equally divided into 14 classes. How many are there in each class?
c In the class library each shelf, on average, holds 18 books. How many shelves will be needed for 432 books?
d Fifteen biscuits make one packet. How many packets can be made from 476 biscuits? How many are left?
e How many perfume bottles, each holding 16 millilitres, can be filled from a ½-litre container?

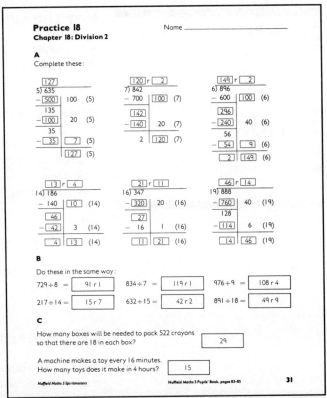

Practice 18
Chapter 18: Division 2   Name _____

A
Complete these:

```
  127
5) 635
− 500  100  (5)
  135
− 100   20  (5)
   35
−  35    7  (5)
       127  (5)
```

```
     120 r  2
7) 842
− 700  100  (7)
  142
− 140   20  (7)
    2  120  (7)
```

```
     149 r  2
6) 896
− 600  100  (6)
  296
− 240   40  (6)
   56
−  54    9  (6)
    2  149  (6)
```

```
      13 r  4
14) 186
− 140   10  (14)
   46
−  42    3  (14)
    4   13  (14)
```

```
      21 r 11
16) 347
− 320   20  (16)
   27
−  16    1  (16)
   11   21  (16)
```

```
      46 r 14
19) 888
− 760   40  (19)
  128
− 114    6  (19)
   14   46  (19)
```

B

Do these in the same way:

729 ÷ 8 = 91 r 1   834 ÷ 7 = 119 r 1   976 ÷ 9 = 108 r 4

217 ÷ 14 = 15 r 7   632 ÷ 15 = 42 r 2   891 ÷ 18 = 49 r 9

C

How many boxes will be needed to pack 522 crayons so that there are 18 in each box?   29

A machine makes a toy every 16 minutes. How many toys does it make in 4 hours?   15

Nuffield Maths 5 Spiritmasters   *Nuffield Maths 5 Pupils' Book, pages 83–85*   **31**

References and resources

Nuffield Mathematics Teaching Project, *Computation and Structure* ⑤
Nuffield Guide, Chambers/Murray 1967 (See Introduction, page xi.)

Williams, E. M. and Shuard, H. *Primary Mathematics Today*, Third
Edition (Chapter 16), Longman Group Ltd 1982

Chapter 19

Time

For the teacher

This chapter introduces the 24-hour clock and deals with the calculation of time intervals and the use of time-tables based on the 24-hour system.

Summary of the stages

1 The 24-hour clock
2 From one time to another
3 Time-tables
4 Multiplication of time

Vocabulary

a.m., p.m., 24-hour system, digital display, depart, arrive, hour(s) (h), minute(s) (min).

Equipment and apparatus

Clock face rubber stamp (with numerals 1 to 12 and 13 to 24), strips of centimetre-squared paper, local bus or train time-tables.

Working with the children

1 The 24-hour clock

Without using a.m. or p.m., the 24-hour system makes it possible to indicate any time of the day unambiguously by using just *four digits and a full stop*.

The first two digits (from 00 to 23) give the number of hours.

The last two digits (from 00 to 59) give the number of minutes.

A full stop separates the hours and minutes.

Translation from a.m./p.m. to the 24-hour system

Range of times	Alteration	Example
All exact hours	.00 required for the minutes	11 a.m. → 11.*00*
Morning times before 10 a.m.	0 required as the first hours digit	7.30 a.m. → *07*.30
Times from 10 a.m. to noon	No alteration (except exact hours)	10.45 a.m. → 10.45 (12 noon → 12.00)
Times between noon and midnight (i.e. p.m.)	Add 12 to the number of hours	5.25 p.m. → *17*.25 9 p.m. → *21.00*★

This last case was explained in a precise, logical but almost matter-of-fact manner by a ten year-old girl as follows:

'Well, p.m. means the last half of the day. Half a day is 12 hours, so you just add 12 to the number of hours.'

* Unfortunately, 21.00 is often called 'twenty-one hundred hours', which is misleading because '00' refers to 'no minutes' or 'no sixtieths of an hour'; there are no 'hundreds' involved.

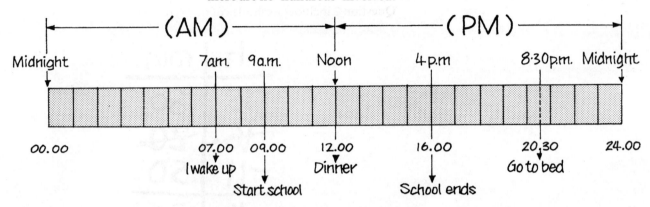

The 'time line' should help children to think of a day as one 24-hour unit rather than as two 12-hour time spans. Fixing times which are of particular interest to children (the match starts at 14.30, school ends at 16.00) will provide 'pegs' on which to hang the 24-hour system. Changing the times on the class time-table to the 24-hour system will provide extra 'pegs'. (*Nuffield Maths 5 Spiritmasters*, Grids 8 and 9.)

Afternoon and early evening times seem to give the most trouble, particularly when translating from the 24-hour to the a.m./p.m. system. 15.00 tends to be thought of as 5 p.m. instead of 3 p.m.; 17.30 as half past 7 instead of half past 5. This is due to subtracting 10 hours instead of 12. The special 'pegs' mentioned earlier should help to avoid these errors.

2 From one time to another

In the a.m./p.m. system, the only direct method for calculating intervals of time which go through noon (that is from a.m. to p.m.) is by counting on from the earlier time to the later. Attempting subtraction by the usual 'on-paper' method leads to complications!

However, in the 24-hour system, the 'on-paper' method of subtraction is straightforward as long as we remember that there are 60 minutes in an hour when decomposition is required. (This is dealt with in the next section.)

$$
\begin{array}{r}
4{:}25\,\text{pm} \\
-\;9.15\,\text{am} \\
\hline
?\,.\,10
\end{array}
$$

h	min
16	25
−09	15
7	10

3   Time-tables

The examples in the first question are chosen so that the number of minutes in the departure time is not greater than the number of minutes in the arrival time so that decomposition of 1 hour into 60 minutes is not necessary.

A worked example shows how to deal with the problem of 'not enough minutes' by exchanging an hour for 60 minutes. 16 h 20 min becomes 15 h 80 min.

Question 2 includes examples of this type.

h	min
15	80
~~16~~	~~20~~
−11	50
4	30

Some children may need help in answering the questions about the train time-table. (Many adults look upon time-tables as mysterious documents!) Extra practice can be given by using part of a local time-table mentioning places familiar to the children. The final question, making up a time-table for trains from Exmouth to Exeter Central, should indicate whether or not children really understand the structure of time-tables.
(*Nuffield Maths 5 Spiritmasters*, Grid 10.)

4   Multiplication of time

Examples are restricted to multiplication of hours and minutes by a single digit. Working in columns headed h and min should remind children that the small '1' under the answer in the 'h' column represents 60 minutes, a small '2' represents 120 minutes, and so on.

h	min
3	35
×	4
14	20
2 ← − − = 140	

Pages from the Pupils' Book and Spiritmasters

86

Chapter 19: Time

The 24-hour clock

The clock shows half-past eight or 8.30. To make sure people know if we mean morning or evening, we have to say a.m. (before noon) or p.m. (after noon).
As there are 24 hours in a day, if we use the 24-hour system for telling the time we do not have to say a.m. or p.m.
8.30 a.m. is 08.30   8.30 p.m. is 20.30
Notice that each number on the inner ring (the p.m. part) is 12 more than the number in the outer ring.
6 a.m. is 06.00   but 6 p.m. is 18.00.
11.15 a.m. is 11.15 but 11.15 p.m. is 23.15.

I Use a strip of squared paper to make a 'time line' like this:

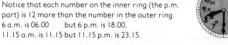
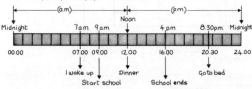

Mark in some special times of the day using both a.m./p.m. and 24-hour systems.

2 Write these times using the 24-hour system. Always use two figures for the hours (01, 02, 03, . . . 09, 10, 11, . . . 23, 24) and two figures for the minutes (00 . . . 59).

a 1 a.m.	**e** 5 p.m.	**i** 10.25 p.m.	**m** 1 minute to midnight
b 1 p.m.	**f** midnight	**j** 8.45 p.m.	**n** 1 minute past midnight
c noon	**g** 9.30 p.m.	**k** 2.35 p.m.	**o** 5 past noon
d 5 a.m.	**h** 10.15 a.m.	**l** 5 past 3 p.m.	**p** 10 to 3 p.m.

87

From one time to another

Working out how long from 9.15 a.m. to 4.25 p.m., for example, is much easier using the 24-hour system.

Using the 12-hour system:

	h	min
from 9.15 a.m. to 10 a.m. is		45
from 10 a.m. to noon is	2	
from noon to 4.25 p.m. is	4	25
	6	70 = 7h 10min.

 from to

Using the 24-hour system

	h	min
	16	25
−	09	15
	7	10

from to

I Use the 24-hour system to find how long between times shown by the two clock faces or digital displays:

a from   to   **b** from   to

a.m.   p.m.   a.m.   p.m.

c from   to   **d** from   to   **e** from   to

2 a Anne starts school at 09.15 and finishes at 15.45. How many hours and minutes is she at school?
b If she has 1¼ hours for dinner time and two play-times of 15 minutes each, how much time is there for lessons?

3 a A marathon race started at 14.05 and the winner crossed the finishing line at 19.53. How long did he take?
b The second runner came in 12 minutes later. At what time did he finish?

88

Train, bus and airway timetables are printed in the 24-hour system. They do not have to print a.m. or p.m. hundreds of times.

I Work out the time taken for each of these journeys:

	depart	arrive		depart	arrive		depart	arrive
a	11.00	15.00	**d**	01.00	18.00	**g**	14.15	19.28
b	08.00	09.30	**e**	05.00	14.30	**h**	09.32	13.48
c	16.00	23.00	**f**	06.30	12.45	**i**	07.10	07.37

If there are 'not enough' minutes for a subtraction, exchange 1 hour for 60 minutes.
16.20 becomes 15h and 80mins.

	15.80
	~~16.20~~
−	11.50
	4.30

2 Change these times to the 24-hour system and find the time taken for each journey:

	depart	arrive		depart	arrive		depart	arrive
a	10.15 a.m.	5 p.m.	**d**	8.35 a.m.	9.10 a.m.	**g**	3.48 a.m.	13.40 p.m.
b	5.30 a.m.	noon	**e**	8.35 a.m.	9.10 p.m.	**h**	5.55 p.m.	9.12 p.m.
c	11.45 a.m.	9.15 p.m.	**f**	6.50 a.m.	1.15 p.m.	**i**	7.32 a.m.	5.10 p.m.

Sometimes the full stop between hours and minutes is missed out, so that a railway timetable looks like this:

Exeter Central	0639	0758	1352	1642	2253
Topsham	0651	0815	1404	1655	2305
Exton	0655	0819	1408	1659	2309
Commando Camp	0657	0821	1410	1701	2311
Exmouth	0704	0828	1417	1708	2318

3 a Do the trains all take the same time over each part of the journey?
b List the time taken from each station to the next by the 2253 train from Exeter Central.
c At what time does the slowest train leave Exeter Central?
d If you arrive at Topsham at half-past three in the afternoon, how long will you have to wait for the next train to Exmouth?
e What is the time of the latest train you can catch from Exeter Central to be sure of getting to Exmouth by 6 p.m.?

4 Make a timetable of your own for journeys from Exmouth to Exeter Central.

89

Multiplication of time

John's father works 7 hours 15 minutes a day.
How many hours and minutes does he work in a 5-day week?

 To multiply 7h 15 min by 5:

Set out the hours and minutes in separate columns like this:

Multiply the 15 min by 5 and write the answer, 75 mins, below the answer line.

75 min = (60 min + 15 min) = 1h 15 min

Write 15 in the minutes column and the small 1 under the hours column.

7h × 5 = 35h. Add on the 1h to make 36h.

	h	min
	7	15
×		5
	36	15
	1	75

I Set these out in the same way:

a 2h 25 min × 3	**d** 3h 10 min × 9	**g** 8h 45 min × 7
b 4h 15 min × 4	**e** 6h 12 min × 5	**h** 4h 52 min × 4
c 8h 5 min × 8	**f** 7h 22 min × 6	**i** 9h 58 min × 8

2 A school's heating is switched on at 08.30 and switched off at 15.10 each day.
a How long is the heating on each day?
b How long is the heating on in a 5-day week?

3 A nurse is on duty from 12.10 to 20.00 each day for 6 days. How many hours is this?

4 The night-duty nurse starts at 20.00 and comes off duty at 06.15 the next day.
a How long is this?
b How many hours duty will a night nurse do in 6 nights?

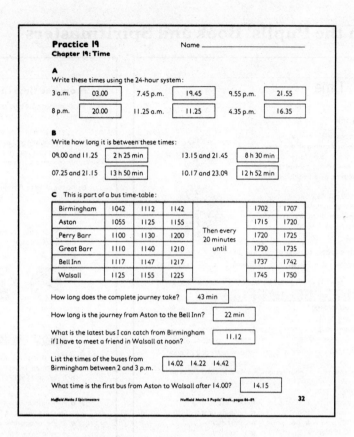

Practice 19
Chapter 19: Time

Name _____

A

Write these times using the 24-hour system:

3 a.m. `03.00`   7.45 p.m. `19.45`   9.55 p.m. `21.55`

8 p.m. `20.00`   11.25 a.m. `11.25`   4.35 p.m. `16.35`

B

Write how long it is between these times:

09.00 and 11.25 `2 h 25 min`   13.15 and 21.45 `8 h 30 min`

07.25 and 21.15 `13 h 50 min`   10.17 and 23.09 `12 h 52 min`

C   This is part of a bus time-table:

Birmingham	1042	1112	1142		1702	1707
Aston	1055	1125	1155		1715	1720
Perry Barr	1100	1130	1200	Then every	1720	1725
Great Barr	1110	1140	1210	20 minutes	1730	1735
Bell Inn	1117	1147	1217	until	1737	1742
Walsall	1125	1155	1225		1745	1750

How long does the complete journey take?   `43 min`

How long is the journey from Aston to the Bell Inn?   `22 min`

What is the latest bus I can catch from Birmingham
if I have to meet a friend in Walsall at noon?   `11.12`

List the times of the buses from
Birmingham between 2 and 3 p.m.   `14.02  14.22  14.42`

What time is the first bus from Aston to Walsall after 14.00?   `14.15`

Nuffield Maths 5 Spiritmasters   *Nuffield Maths 5 Pupils' Book, pages 86–89*   **32**

References and resources

Science 5–13 Project, *Time. A unit for teachers*, Macdonald 1972

Arnold, E. J. *24-hour clock face rubber stamp*

Julie Bloomfield & Associates, 42 Tavistock Street, WC2E 7PB, *Timex Posters* (9 posters and teachers' notes)

Philip & Tacey Ltd, *24-hour clock face rubber stamp*

Chapter 20

Fractions 2

For the teacher

The work in Chapter 14 is extended here to include mixed numbers – that is numbers which are a mixture of whole numbers and fractional parts; for example:

$$1\frac{3}{4} \text{ or } 7\frac{5}{6}.$$

This work should not be rushed and diagrams and apparatus should be used by those who find the work difficult.

Summary of the stages

1 Mixed and improper fractions
2 Addition of mixed numbers
3 Subtraction of mixed numbers

Vocabulary

Mixed numbers, improper fractions.

Equipment and apparatus

Squared paper.

Working with the chidren

1 Mixed and improper fractions

There are two ways of counting in fifths:

(i) $\dfrac{1}{5}, \dfrac{2}{5}, \dfrac{3}{5}, \dfrac{4}{5}, \dfrac{5}{5}, \dfrac{6}{5}, \dfrac{7}{5}, \dfrac{8}{5} \cdots\cdots$

Here when we get beyond one $\left(\dfrac{5}{5}\right)$ we continue to count in fifths

and these are *improper fractions*. They are easily recognised as their numerators are larger than their denominators.

(ii) $\dfrac{1}{5}, \dfrac{2}{5}, \dfrac{3}{5}, \dfrac{4}{5}, 1, 1\dfrac{1}{5}, 1\dfrac{2}{5}, 1\dfrac{3}{5} \cdots\cdots$

In this way of counting one is called 'one' and not 'five fifths' and *mixed numbers* – a mixture of whole numbers and fractional parts – follow. These two ways are shown on this diagram:

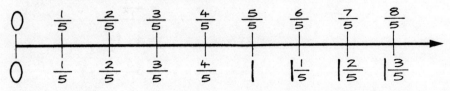

As can be readily seen they are different ways of writing the same thing – for example,

$$\frac{7}{5} = 1\frac{2}{5}.$$

Changing from one to the other is an important part of fraction work and this can be introduced to the pupils by diagrams or using apparatus.

This diagram shows the mixed number $2\frac{5}{6}$. Each whole one is $\frac{6}{6}$ so altogether

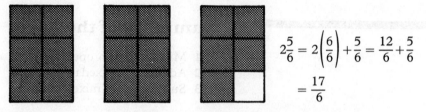

$$2\frac{5}{6} = 2\left(\frac{6}{6}\right) + \frac{5}{6} = \frac{12}{6} + \frac{5}{6}$$
$$= \frac{17}{6}$$

This diagram shows the improper fraction $\frac{9}{4}$.

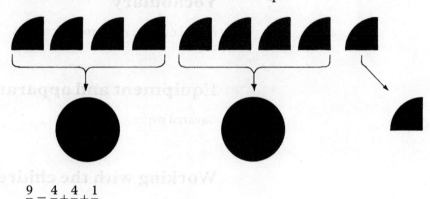

$$\frac{9}{4} = \frac{4}{4} + \frac{4}{4} + \frac{1}{4}$$
$$= 2\frac{1}{4}$$

2 Addition of mixed numbers

When adding mixed numbers, for example $2\frac{1}{2} + 1\frac{1}{3}$ the whole numbers are added first. This is approached through the 'jars of jam' example in the *Pupils' Book*.

The first question produces mixed number answers but the second presents the difficulty of improper fractions in the answer. The method given in the *Pupils' Book* is:

$$3\frac{2}{3} + 1\frac{3}{4} = 4 + \frac{8}{12} + \frac{9}{12}$$
$$= 4\frac{17}{12}$$
$$= 4 + 1\frac{5}{12}$$
$$= 5\frac{5}{12}$$

3 Subtraction of mixed numbers

When subtracting mixed numbers it is sometimes necessary to decompose a whole into fractional parts. This is introduced to the children diagrammatically.

$$2\frac{3}{6} - \frac{4}{6}$$

$$1\frac{9}{6} - \frac{4}{6}$$

Changing 1 whole into 6 sixths in this example follows the same technique as changing 1 ten into 10 units when using decomposition for the subtraction of whole numbers.

$$\begin{array}{r} {}^{2}{}^{14} \\ \cancel{3}\cancel{4} \\ -\ 18 \\ \hline \end{array}$$

Here is a further example with the explanation for each step:

$$5\frac{1}{4} - 1\frac{3}{5}$$ Subtract the whole numbers

$$= 4\frac{1}{4} - \frac{3}{5}$$ Change the fractions into twentieths

$$= 4\frac{5}{20} - \frac{12}{20}$$ There are not enough twentieths to take away 12 so change one of the wholes into twentieths.

$$= 3\frac{25}{20} - \frac{12}{20}$$

$$= 3\frac{13}{20}$$

Pages from the Pupils' Book and Spiritmasters

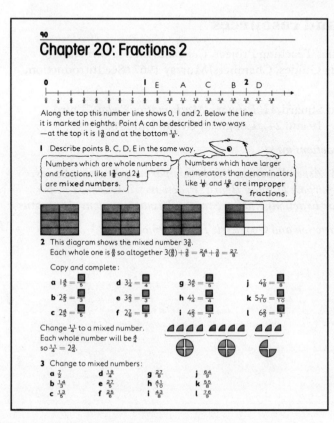

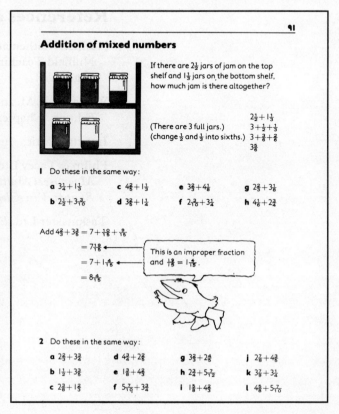

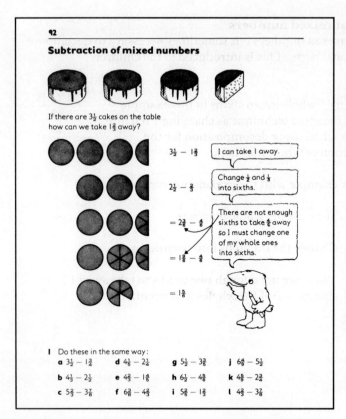

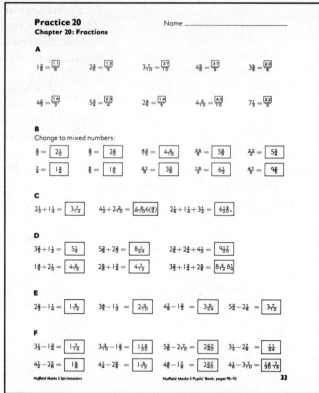

References and resources

Nuffield Mathematics Teaching Project, *Computation and Structure* ⑤ Nuffield Teaching Guides, Chambers/Murray 1967 (See Introduction, page xi.)

Williams, E. M. and Shuard, H. *Primary Mathematics Today*, Third Edition (Chapters 19 and 24), Longman Group Ltd 1982

Invicta Plastics, *Fractions kit*

Philip & Tacey Ltd, *Aspex Fractions Set 1, Comparative Fraction Strips, Magna-cel Multipurpose Display Board, Practi-metric Gummed Paper Shapes, Visi-Clear Fractions Rubber Stamps, Visual Fractions Apparatus*

Taskmaster Ltd, *Fraction and Geometric Pieces, Pinboards*

Chapter 21

Length 2

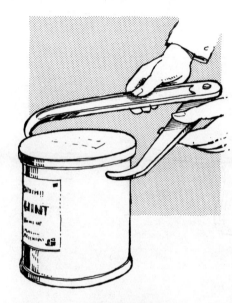

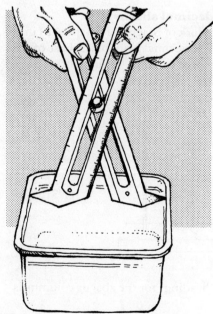

For the teacher

The introduction of the millimetre (mm) enables earlier activities involving the measurement and calculation of length to be repeated but to a greater degree of accuracy. Some children may not be ready to deal with units as small as millimetres but the use of a magnifying glass should make it possible for most. In the same way that decimetres and centimetres helped with 'the bit left over' when measuring in metres, so millimetres make more accurate measurement of shorter lengths possible. Even so, measurement can never be exact, and children may be interested to know that even smaller units such as the micrometre (0.000001 m) exist.

Summary of the stages

1 Measuring in millimetres
2 Recording measurements on a decimal abacus
3 Calculation in millimetres
4 Using 2-mm graph paper

Vocabulary

Millimetre(s) (mm), centimetre(s) (cm), decimetre(s) (dm), metre(s) (m), perimeter, regular octagon, regular hexagon, diameter, estimate.

Equipment and apparatus

Rulers marked in centimetres and millimetres, callipers, column abaci, magnifying glass, graph paper marked in 2-millimetre squares.

Working with the children

1 Measuring in millimetres

The exercises should be supplemented with additional objects which have clearly defined end-points. Objects with curved ends, diameters of small circular objects or internal measurements are best dealt with using callipers.

The distance between the points of the callipers is set to represent the length being measured. This enables the length to be transferred to and 'read off' from a ruler marked in millimetres.

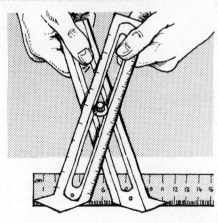

Some callipers are graduated so that a direct reading of the length can be made from the scale.

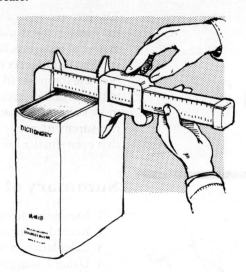

For very small diameters or thicknesses a simple micrometer with widely spaced graduations can be used.

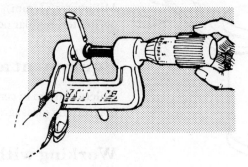

2 Recording measurements on a decimal abacus

This section reinforces and extends the work of Stage 3 of Chapter 13 in *Nuffield Maths 4 Teachers' Handbook*. The decimal point of the base of the abacus must be moved so that there are three columns to the right. (*Nuffield maths 5 Spiritmasters*, Grid 30.)

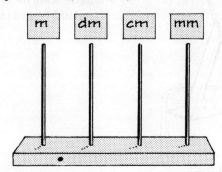

As before, the gradual change of the headings for the abacus columns is left to the discretion of the teacher.

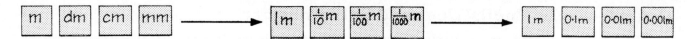

The relationship between mm and cm, cm and dm, dm and m illustrates the 'ten-ness' of the system and reinforces *place value*. Previous work on *equivalent fractions* will also be revised.

For example, $7 \text{ cm} \left(\frac{7}{100} \text{ m} \right) = 70 \text{ mm} \left(\frac{70}{1000} \text{ m} \right)$

Teachers may find it useful to display the 'system' in a tabular form:

1 metre = 10 decimetres	= 100 centimetres	= 1000 millimetres
1 decimetre	= 10 centimetres	= 100 millimetres
1 centimetre	=	10 millimetres

Or, reversing the order:

$$1 \text{ mm} = \frac{1}{10} \text{ cm} = \frac{1}{100} \text{ dm} = \frac{1}{1000} \text{ m}$$
$$1 \text{ cm} = \frac{1}{10} \text{ dm} = \frac{1}{100} \text{ m}$$
$$1 \text{ dm} = \frac{1}{10} \text{ m}$$

Or, using decimal notation:

$$1 \text{ mm} = 0.1 \text{ cm} = 0.01 \text{ dm} = 0.001 \text{ m}$$
$$1 \text{ cm} = 0.1 \text{ dm} = 0.01 \text{ m}$$
$$1 \text{ dm} = 0.1 \text{ m}$$

Again, it is strongly recommended that 'short cut tricks' such as 'putting in the decimal point' are *not* suggested to the child. It is far more important at this stage for children to see *why* figures appear in their respective columns than to acquire a slick technique.

Recording lengths such as 5 mm or 23 mm on the decimal abacus first gives a visual reminder of the need for correct placing of zeros in the written version. (*Nuffield Maths 5 Spiritmasters*, Grid 30.)

3 Calculation in millimetres
Questions on the measurement and calculation of perimeters lead to addition, subtraction, multiplication and division (by a single digit) of lengths including millimetres. The format of this section is similar to that used in Chapter 8, Length 1.

4 Using 2-mm graph paper
This short section gives further experience of accurate measurement in mm and suggests the drawing of graphs. If 2-mm squared paper is used for the graphs, dimensions of smaller objects can be transferred directly on to the paper without having to select a suitable scale. This in turn gives practice in accurate drawing and the interpretation of graphs.

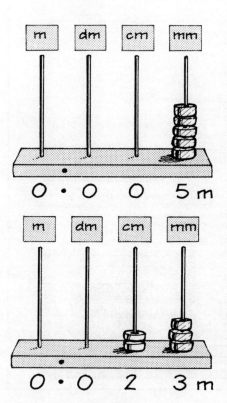

Pages from the Pupils' Book and Spiritmasters

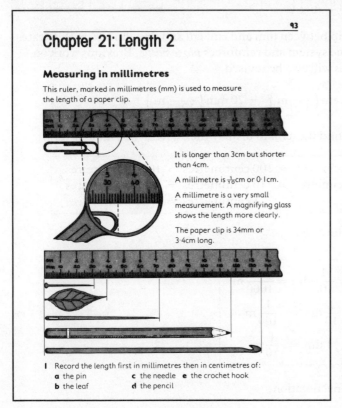

Chapter 21: Length 2

Measuring in millimetres

This ruler, marked in millimetres (mm) is used to measure the length of a paper clip.

It is longer than 3cm but shorter than 4cm.

A millimetre is $\frac{1}{10}$cm or 0·1cm.

A millimetre is a very small measurement. A magnifying glass shows the length more clearly.

The paper clip is 34mm or 3·4cm long.

1 Record the length first in millimetres then in centimetres of:
a the pin   c the needle   e the crochet hook
b the leaf   d the pencil

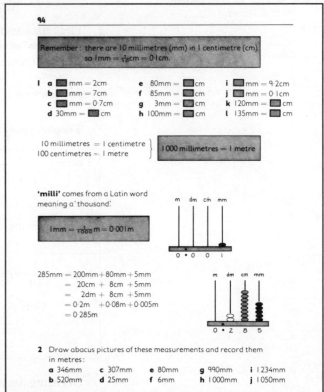

Remember: there are 10 millimetres (mm) in 1 centimetre (cm) so 1mm = $\frac{1}{10}$cm = 0·1cm.

1 a ▓mm = 2cm   e 80mm = ▓cm   i ▓mm = 9·2cm
   b ▓mm = 7cm   f 85mm = ▓cm   j ▓mm = 0·1cm
   c ▓mm = 0·7cm   g 3mm = ▓cm   k 120mm = ▓cm
   d 30mm = ▓cm   h 100mm = ▓cm   l 135mm = ▓cm

10 millimetres = 1 centimetre
100 centimetres = 1 metre   } 1000 millimetres = 1 metre

'milli' comes from a Latin word meaning a 'thousand'.

1mm = $\frac{1}{1000}$m = 0·001m

285mm = 200mm + 80mm + 5mm
      = 20cm + 8cm + 5mm
      =   2dm + 8cm + 5mm
      = 0·2m + 0·08m + 0·005m
      = 0·285m

2 Draw abacus pictures of these measurements and record them in metres:
   a 346mm   c 307mm   e 80mm   g 990mm   i 1234mm
   b 520mm   d 25mm   f 6mm   h 1000mm j 1050mm

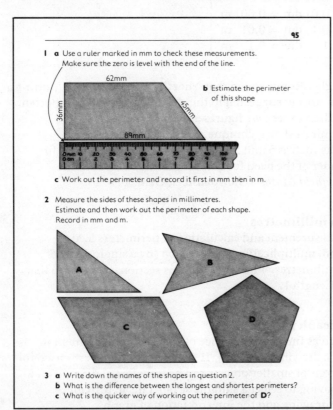

1 a Use a ruler marked in mm to check these measurements. Make sure the zero is level with the end of the line.

62mm
36mm
45mm
89mm

b Estimate the perimeter of this shape

c Work out the perimeter and record it first in mm then in m.

2 Measure the sides of these shapes in millimetres. Estimate and then work out the perimeter of each shape. Record in mm and m.

A   B   C   D

3 a Write down the names of the shapes in question 2.
   b What is the difference between the longest and shortest perimeters?
   c What is the quicker way of working out the perimeter of D?

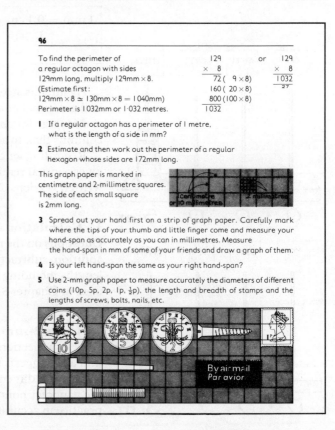

To find the perimeter of a regular octagon with sides 129mm long, multiply 129mm × 8.
(Estimate first: 129mm × 8 ≃ 130mm × 8 = 1040mm)
Perimeter is 1032mm or 1·032 metres.

```
  129        or    129
×   8            ×   8
 72 ( 9 × 8)     1032
160 (20 × 8)       27
800 (100 × 8)
1032
```

1 If a regular octagon has a perimeter of 1 metre, what is the length of a side in mm?

2 Estimate and then work out the perimeter of a regular hexagon whose sides are 172mm long.

This graph paper is marked in centimetre and 2-millimetre squares. The side of each small square is 2mm long.

3 Spread out your hand first on a strip of graph paper. Carefully mark where the tips of your thumb and little finger come and measure your hand-span as accurately as you can in millimetres. Measure the hand-span in mm of some of your friends and draw a graph of them.

4 Is your left hand-span the same as your right hand-span?

5 Use 2-mm graph paper to measure accurately the diameters of different coins (10p, 5p, 2p, 1p, $\frac{1}{2}$p), the length and breadth of stamps and the lengths of screws, bolts, nails, etc.

By air mail
Par avior

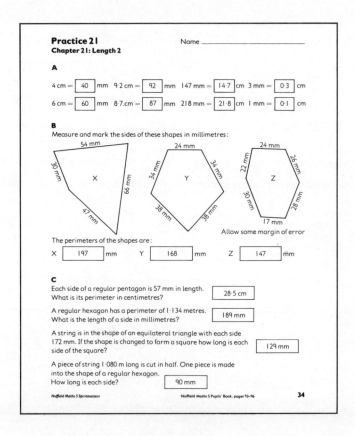

Practice 21
Chapter 21: Length 2

Name _____

A

4 cm = $\boxed{40}$ mm   9·2 cm = $\boxed{92}$ mm   147 mm = $\boxed{14·7}$ cm   3 mm = $\boxed{0·3}$ cm

6 cm = $\boxed{60}$ mm   8·7 cm = $\boxed{87}$ mm   218 mm = $\boxed{21·8}$ cm   1 mm = $\boxed{0·1}$ cm

B
Measure and mark the sides of these shapes in millimetres:

Allow some margin of error

The perimeters of the shapes are:

X $\boxed{197}$ mm   Y $\boxed{168}$ mm   Z $\boxed{147}$ mm

C

Each side of a regular pentagon is 57 mm in length.
What is its perimeter in centimetres? $\boxed{28·5 \text{ cm}}$

A regular hexagon has a perimeter of 1·134 metres.
What is the length of a side in millimetres? $\boxed{189 \text{ mm}}$

A string is in the shape of an equilateral triangle with each side
172 mm. If the shape is changed to form a square how long is each
side of the square? $\boxed{129 \text{ mm}}$

A piece of string 1·080 m long is cut in half. One piece is made
into the shape of a regular hexagon.
How long is each side? $\boxed{90 \text{ mm}}$

Nuffield Maths 5 Spiritmasters   *Nuffield Maths 5 Pupils' Book, pages 93–96*   **34**

References and resources

Nuffield Maths Teaching Project, *Computation and Structure* ⑤, Nuffield
  Guide, Chambers/Murray 1967 (See Introduction, page xi.)

Williams, E. M. and Shuard, H. *Primary Mathematics Today*, Third
  Edition (Chapter 24), Longman Group Ltd 1982

Arnold, E. J. *Abacus Board and tablets, Harex Abacus*

Classmate Triman, *Decimal Abacus, Combination Set*

E.S.A. *Base 10 Abacus*

Invicta Plastics, *Abacus and tablets*

Osmiroid, *Bow Callipers, Graduated Callipers, Micrometer*

Answers to Nuffield Maths 5 Pupils' Book

Chapter 1: Addition 1

Page 2

1a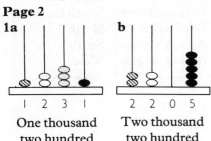

One thousand
two hundred
and thirty-one

b
Two thousand
two hundred
and five

c 2110
Two thousand
one hundred
and ten

d 1032
One thousand
and thirty-two

2a 5143 Five thousand one
hundred and forty-three

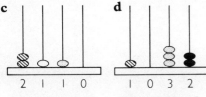

b 2305 Two thousand three
hundred and five

c 3017 Three thousand and
seventeen

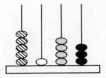

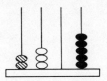

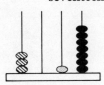

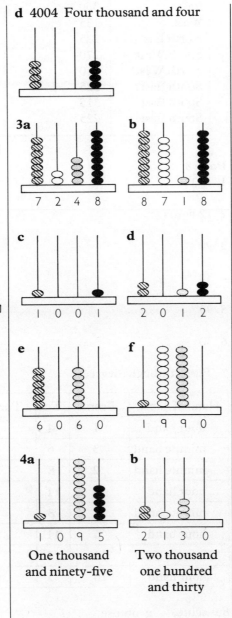

d 4004 Four thousand and four

3a 7248

b 8718

c 1001

d 2012

e 6060

f 1990

4a 1095
One thousand
and ninety-five

b 2130
Two thousand
one hundred
and thirty

c 5040
Five thousand
and forty

d 3993
Three thousand
nine hundred
and ninety-three

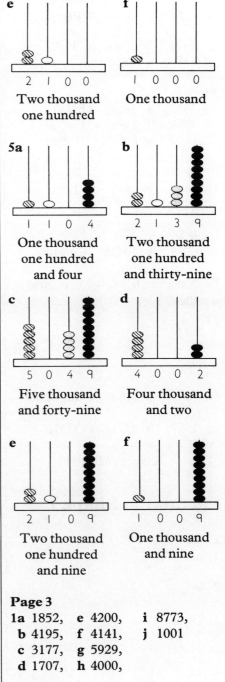

e 2100
Two thousand
one hundred

f 1000
One thousand

5a 1104
One thousand
one hundred
and four

b 2139
Two thousand
one hundred
and thirty-nine

c 5049
Five thousand
and forty-nine

d 4002
Four thousand
and two

e 2109
Two thousand
one hundred
and nine

f 1009
One thousand
and nine

Page 3

1a 1852, **e** 4200, **i** 8773,
b 4195, **f** 4141, **j** 1001
c 3177, **g** 5929,
d 1707, **h** 4000,

Page 4

1a   2174   **b**   718
      +3723   +2131
       5897   2849

c 1357	**d** 5665
+8642	+2024
9999	7689

2a 1575   **b** 4189,
  +2319   **c** 7205,
  3894   **d** 6219

3a 4954, **b** 5580, **c** 4570,
d 3390, **e** 4300

Page 5
1a 3682, **e** 4956, **h** 1411,
b 4767, **f** 2798, **i** 3920,
c 7170, **g** 699,   **j** 7064
d 8000,

2a 7791, **b** 1847, **c** 1255,
d 2448

Page 6
1a 6688, **b** 10520, **c** 8914

2a 2000,   **b** 1800,   **c** 2055,
d 5855

3 2044

4 1901

5a 5204, **b** 6256, **c** 1880 km

6a 1389, **b** 3371, **c** 3291,
d 4975, **e** 3076,
 f Both should equal 8051

Chapter 2: Shape 1

Page 7
1 90°

2 45°

3 135°

4

I start facing	I turn through	Direction	I finish facing
North	180°	Clockwise	South
South	90°	Anti-clockwise	East
East	360°	Clockwise	East
West	135°	Clockwise	North East
North East	225°	Anti-clockwise	South
South West	270°	Clockwise	South East
North West	270°	Anti-clockwise	North East
South East	225°	Clockwise	North
South East	315°	Anti-clockwise	South
South East	225°	Anti-clockwise	West

Page 8
1 360°

2 12 hours

3 90°

4 30°

5 90°

6 30°

7 Turns through (degrees)

Hand of Clock	From	To	Turns through°
hour hand	1	4	90°
minute hand	3	6	90°
minute hand	2	8	180°
hour hand	5	11	180°
minute hand	4	8	120°
hour hand	6	11	150°
minute hand	11	4	150°
hour hand	10	5	210°

8a acute,   **g** obtuse,
 b obtuse, **h** obtuse,
 c acute,   **i** acute,
 d acute,   **j** obtuse,
 e obtuse, **k** acute,
 f acute,   **l** obtuse

Page 9
1 Activity

2a 50°, **b** 60°, **c** 60°,
 d 50°, **e** 60°, **f** 59°

Page 10
1 60°

2 55° and 70°

3 90°

4 45°

5 45°

6 Right-angled; isosceles

7 **g** 30°, **h** 60°, **i** 60°, **j** 60°,
 k 40°, **m** 40°, **n** 70°

Chapter 3: Subtraction 1

Page 12
1a 1323,   **c** 776,
 b 1736,   **d** 1655

2 4000 = 3(1000) + 9(100)
                 + 9(10) + 10(1)

3a 520,   **f** 1688,
 b 2288,   **g** 1879,
 c 71,   **h** 4731,
 d 1948,   **i** 3087,
 e 2655,   **j** 1278

4a 591, **b** 56, **c** 219,
 d 372

5 The highest score is 647 greater
  than the lowest score.

Page 13
Red must win 275 pts.
1a 128 + [39] = 167
 b 1286 + [124] = 1410
 c 1679 + [356] = 2035
 d 2345 + [1858] = 4203
 e 4298 + [2073] = 6371
 f 5920 + [2719] = 8009

2a 32,   **c** 210,   **e** 216,
 b 45,   **d** 333,   **f** 1334

Page 14
1a 883,   **d** 883,   **f** 700,
 b 789,   **e** 139,   **g** 233,
 c 789

2 257

3 595 cm³

4a 396 paces, **b** 3960 paces

5a Plantagenets, **b** Normans,
 c 213 years more,
 d 99 years shorter.

6 239 km

7 £2757

Chapter 4: Area 1

Page 15

1	Length in centimetres	Breadth in centimetres	Area in square centimetres	Perimeter in centimetres
A	6 cm	4 cm	24 cm²	20 cm
B	8 cm	3 cm	24 cm²	22 cm
C	12 cm	2 cm	24 cm²	28 cm

2 The areas of A, B and C are
  equal, but they have different
  perimeters.

3a Area 21 cm², perimeter 20 cm
 b Area 9 cm², perimeter 20 cm
 c Breadth 2 cm, perimeter 20 cm
 d Area 25 cm²

4 The areas of D, E, F and G are all
  different but they have equal
  perimeters.

Page 16
1 Activity
2a 100 cm²
 b Rectangle 50 cm
   Triangle 48 cm
   Boat 48 cm

c The shapes have the same area,
  the triangle and the boat have
  the same perimeter, but the
  perimeter of the rectangle is
  longer.

3 Measuring activity

Page 17
1, 2 Open answers

Page 18
1a 30 dm², 36 dm²
 b 3000 cm², 3600 cm²
 c Picnic table has the greater area
   by 600 cm²
 d 260 cm, 240 cm
 e The coffee table has a smaller
   area than the picnic table but a
   greater perimeter.

2	Shape	Number of dm² squares to cover	Area in sq cm	Perimeter in cm
	A	18	1800 cm²	200 cm
	B	17	1700 cm²	200 cm
	C	10	1000 cm²	200 cm

Page 19

1 The string will be 4 m long.

2a Carpet: area 58 m², perimeter 36 m
Lawn: area 68 m², perimeter 36 m
The perimeters of the lawn and carpet are the same but they have different ares.

b £464, **c** No, **d** 18 minutes

Chapter 5:
Multiplication 1

Page 20

1a 52,   **b** 112,   **c** 108,
d 153,   **e** 152

2a 72,   **e** 272,   **i** 413,
b 124,   **f** 414,   **j** 282
c 168,   **g** 285,
d 259,   **h** 384,

3a 258,   **b** 296,   **c** 441,
d 486,   **e** 380

Page 21

1a 58 × 5

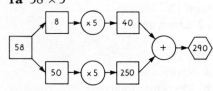

b 75 × 9

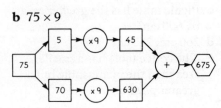

2a

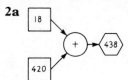

b

c

d

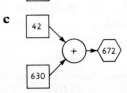

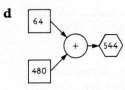

Page 22

1a 81,   **e** 148,   **i** 581,
b 204,   **f** 399,   **j** 828
c 190   **g** 504,
d 258,   **h** 294,

Page 23

1a 402,   **d** 1968,   **g** 3381,
b 1125,   **e** 5160,   **h** 5643
c 704   **f** 3564,

Page 24

1a 112,   **b** 126,
c £1.44 or 144p,   **d** 90,
e 144,   **f** 296,   **g** 216,
h 600 metres,   **i** 324,
j £2.40 or 240p,   **k** 306 cm,
l 288,   **m** 252 ml

Chapter 6:
Volume and Capacity

Page 25

1	Prism	Shape of 'slice'
	B	Square
	C	triangle
	D	hexagon
	G	rectangle

2 **P** 9 cm³ **Q** 16 cm³, **R** 24 cm³
S 20 cm³, **T** 60 cm³

Page 26

1 **A** 12 cm³,   **D** 20 cm³,
B 36 cm³,   **E** 24 cm³,
C 24 cm³,   **F** 30 cm³

2 Cuboids **C** and **E** are of the same volume.

Page 27

Cuboid letter	Area of cross-section	Length	Volume
A	8 cm²	3 cm	24 cm³
B	16 cm²	1 cm	16 cm³
C	1 cm²	4 cm	4 cm³
D	4 cm²	7 cm	28 cm³
E	2 cm²	8 cm	16 cm³

Page 28

1 **A** 45 cm³,   **D** 48 cm³,
   B 24 cm³,   **E** 18 cm³,
   C 42 cm³,   **F** 30 cm³

2

Area of end face square units	Length of prism length units	Volume of prism cubic units
15 cm²	6 cm	90 cm³
22 cm²	9 cm	198 cm³
100 cm²	10 cm	1000 cm³
12 cm²	9 cm	108 cm³
6 cm²	1.50 m	900 cm³

Page 29

1 Area 25 cm², volume 300 cm³.
   Area 48 cm², volume 432 cm³.
   Area 117 cm², volume 819 cm³.

2, 3 Activities

Page 30

1 1000 cm²

2a 16000 cm³,   **b** 16 litres

3a 11000 cm³,   **b** 11 litres

4 13 cm

5 Open answers

Chapter 7: Division 1

Page 31

1a 14 r 3,   **e** 11 r 2,
b 11 r 4,   **f** 13 r 1
c 13 r 2,   **g** 12 r 3,
d 11 r 5,   **h** 13 r 2

2a
```
    23 r 3
4)95
  −40  | 10 (4)
   55
  −40  | 10 (4)
   15
  −12  |  3 (4)
    3  | 23 (4)
```

b
```
    28 r 1
2)57
  −20  | 10 (2)
   37
  −20  | 10 (2)
   17
  −16  |  8 (2)
    1  | 28 (2)
```

c
```
    22 r 2
3)68
  −30  | 10 (3)
   38
  −30  | 10 (3)
    8
  − 6  |  2 (3)
    2  | 22 (2)
```

Page 32

1a
```
    34 r 1
7)239
  −70  | 10 (7)
  169
  −70  | 10 (7)
   99
  −70  | 10 (7)
   29
  −28  |  4 (7)
    1  | 34 (7)
```

b
```
    40
8)320
  −80  | 10 (8)
  240
  −80  | 10 (8)
  160
  −80  | 10 (8)
   80
  −80  | 10 (8)
    0  | 40 (8)
```

c
```
    31 r 4
9)283
  −90  | 10 (9)
  193
  −90  | 10 (9)
  103
  −90  | 10 (9)
   13
  − 9  |  1 (9)
    4  | 31 (9)
```

2a 31 r 3,   **f** 32 r 3,
 b 38 r 3,   **g** 39,
 c 40,   **h** 38 r 3,
 d 32 r 5,   **i** 34,
 e 32,   **j** 36 weeks

Page 33

1a 90,   **d** 190,   **g** 480,
 b 110,   **e** 230,   **h** 730
 c 140,   **f** 340,

2a 180,   **d** 1350, **g** 4760,
 b 320,   **e** 1020, **h** 7110
 c 720,   **f** 4240,

Page 34

1a
```
     43 r 2
8)346
 −320 | 40 (8)
   26
 − 24 |  3 (8)
    2 | 43 (8)
```

b
```
     51
7)357
 −350 | 50 (7)
    7
 −  7 |  1 (7)
    0 | 51 (7)
```

c
```
     71 r 6
9)645
 −630 | 70 (9)
   15
 −  9 |  1 (9)
    6 | 71 (9)
```

d 86 r 1,   **i** 58,
e 44 r 3,   **j** 80 r 7,
f 62,   **k** 91 r 7,
g 64 r 4,   **l** 92 r 2,
h 87,   **m** 52 r 1.

2a 42 trees,   **f** 15 toys,
 b £56,   **g** 84 boxes,
 c 97 cars,   **h** 18 cm,
 d 43 cm,   **i** 23 pens,
 e 57 spoonfuls, **j** 24 marbles.

Chapter 8: Length 1

Page 35

1 A

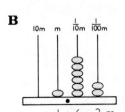

    1 · 3 9 m

B

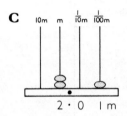

    1 · 6 2 m

C

    2 · 0 1 m

2 39 cm,

3 23 cm,

4 42 cm

Page 36

1a 162 cm, **b** 249 cm, **c** 201 cm

2 Open answers.

3 A 2·35 m, B 4·32 m,
   C 2·64 m.

4 'penta' 5, 'hexa' 6, 'octa' 8.

Page 37

1a They are not regular because
   they do not have equal angles.
 b Yes, they are symmetrical.
 c P 3·48 m, Q 6·79 m,
    R 3·55 m

2 15·75 m

3 12·84 m

Page 38

1a 31 cm, 0·31 m
 b 72 cm, 0·72 m
 c 191 cm, 1·91 m
 d 101 cm, 1·01 m
 e 142 cm, 1·42 m

2 A 52 cm, 0·52 m
   B 81 cm, 0·81 m
   C 91 cm, 0·91 m
   D 140 cm, 1·40 m
   E 131 cm, 1·31 m
   F 106 cm, 1·06 m

Page 39

1 Each skipping rope is 2·76 m or
  276 cm

2 2·30 m or 230 cm

3a 1·49 m, **c** 1·59 m, **e** 3·22 m,
 b 2·67 m, **d** 2·75 m, **f** 2·6 m

4a 7·50 m, **b** 1·50 m, **c** 2·50 m
 d The sides of the triangle are
    twice as long as the sides of the
    hexagon.

Chapter 9: Money

Page 40

1a £4·35, **d** £3·06, **g** £0·73,
 b £6·03, **e** £8·91, **h** £0·92
 c £5·74, **f** £10·24,

2a 173 p, **d** 894 p, **g** 61 p,
 b 349 p, **e** 937 p, **h** 17 p
 c 637 p, **f** 1061 p,

3a £9·00, **c** £16·63, **e** £11·51,
 b £12·92, **d** £13·07, **f** £13·01½

4a £2·51, **c** £2·06, **e** £1·86,
 b £2·53½, **d** £2·03½, **f** £3·22½

5a £3·96, **f** £24·78, **k** £17·95,
 b £3·48, **g** £17·52, **l** Yes,
 c £17·55, **h** £20·64, **m** £44·45
 d £4·56, **i** £52·83,
 e £38·24, **j** £34·72,

Page 41

1a £1·75,
 b £1·39,
 c £1·59,
 d £2·41 r 2,
 e £1·37 r 1,
 f £1·74 r 3,
 g £1·52 r 2,
 h £0·90 r 6,

 i £1·12 r 4,
 j £1·18,
 k £1·13 r 2,
 l £1·55 r 7,
 m £2·45 each,
 n £6·20 per day,
 o £6·95,
 p £2·45 per hour.

Page 42

1a £12·05, b £25·29, c £16·53

2a Shears, by £5·26,
 b Sports bag, by £0·70 or 70 p,
 c Wheel barrow by £8,
 d Lawn edger, by £0·30 or 30 p.

3a £37·40, c £26·94, e £26·32,
 b £16·35, d £34·95, f £59·85

Page 43

1a

	£
4 spades	37·40
3 forks	+ 29·07
5 lawn edgers	32·75
Total	99·22

b

	£
5 table tennis sets	23·75
7 badminton racquets	+ 48·93
4 sports bags	21·80
	94·48

c

	£
8 torches	14·80
6 hoses	+ 26·94
5 watering cans	17·45
Total	59·19

d

	£
3 wheelbarrows	44·97
6 shears	+ 51·30
5 rakes	31·25
	127·52

2 £1·16

3a £1·60,   c £4·99,
 b £2·79,   d £6·47

4 torch 25 p
 vacuum flask 50 p
 sports bag 46 p
 badminton racquet 52 p dearer

5

5 torches	9·25
4 vacuum flasks	13·16
2 sports bags	10·90
3 badminton racquets	20·97
	54·28

This is £5·73 more this year.

Chapter 10: Rounding off

Page 44

1 A $\simeq$ 1   D $\simeq$ 3   G $\simeq$ 4
 B $\simeq$ 1   E $\simeq$ 3   H $\simeq$ 5
 C $\simeq$ 2   F $\simeq$ 4   I $\simeq$ 5

Page 45

1 Line **CD** $\simeq$ 6 cm,
 EF $\simeq$ 5 cm,   **GH** $\simeq$ 4 cm,
 IJ $\simeq$ 8 cm,   **KL** $\simeq$ 3 cm,
 MN $\simeq$ 5 cm,   **OP** $\simeq$ 7 cm,
 QR $\simeq$ 4 cm,   **ST** $\simeq$ 7 cm
2 25 cm,   19 cm

Page 46

1 **A** $\simeq$ 400 ml, **B** $\simeq$ 100 ml,
 C $\simeq$ 400 ml, **D** $\simeq$ 200 ml,
 E $\simeq$ 200 ml

2 Clock **P** a) 4.00,
         b) 3.30,
         c) $\frac{1}{4}$ to 4,
         d) 3.40

 Clock **Q** a) 2.00,
         b) 2.30,
         c) 2.30,
         d) 2.25

 Clock **R** a) 12.00,
         b) 12.00,
         c) 12.15,
         d) 12.15

 Clock **S** a) 9.00,
         b) 9.00,
         c) 9.00,
         d) 9.05

Page 47

1a 3·0,   e 8·0,   i 7·0,
 b 4·0,   f 3·0,   j 4·0
 c 27·0,   g 10·0,
 d 1·0,   h 100·0,

2 50 cm

3a 3 m,   d 6 m,   g 13 m,
 b 8 m,   e 4 m,   h 30 m
 c 7 m,   f 1 m,

4a 15·38 m, b 15 m, c 16 m,
 d Ans c is one metre more than answer b because for c three lengths have been rounded up whereas b has only one length rounded up.

5 Activity

Page 48

1 **A** $\simeq$ 4 cm²,
 B $\simeq$ 8 cm²,
 C $\simeq$ 7 cm²,
 D $\simeq$ 17 cm²

2 7·75 cm

3a No, b 20 cm², c £190

4 Activities

Page 49

1a 8 cm³,   d 216 cm³,
 b 27 cm³,   e 125 cm³
 c 1000 cm³,

2 **A** $\simeq$ 16 cm³,
 B $\simeq$ 18 cm³,
 C $\simeq$ 10 cm³

3 Measuring activity.

Chapter 11: Area 2

Page 50

1a 18 cm², **b** 9 cm²,

c The area of each triangle is half the area of the rectangle.

2 **A** 6 cm², **D** 5 cm²,
B 15 cm², **E** 10 cm²,
C 10·5 cm², **F** 9 cm²

Page 51

1 **A** 6 cm², **D** 15 cm²,
B 10 cm², **E** 22·5 cm²,
C 8 cm², **F** 4·5 cm²

Page 52

1 **A** 10 cm², **C** 17·5 cm²
B 6 cm²,

Page 53

1 **A** 12 cm², **D** 24 cm²,
B 22 cm², **E** 19·5 cm²
C 25 cm²,

2 48 m²

3 30 m²

Chapter 12: Co-ordinates

Page 54

1

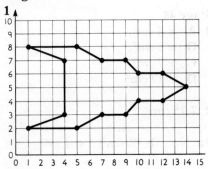

2

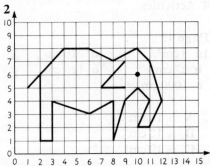

3 Activity

Page 55

1a (0,10) (3,7) (6,4) (9,1)
(1,9) (4,6) (7,3) (10,0)
(2,8) (5,5) (8,2)

1b, c, d and **e** Activities

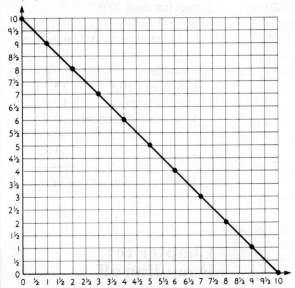

f $(7\frac{1}{2},2\frac{1}{2})$; $(6\frac{1}{2},3\frac{1}{2})$; $(1\frac{1}{2},8\frac{1}{2})$;
$(5\frac{1}{2},4\frac{1}{2})$; $(\frac{1}{2},9\frac{1}{2})$; $(8\frac{1}{2},1\frac{1}{2})$;
$(5\frac{1}{2},4\frac{1}{2})$

Page 56

1

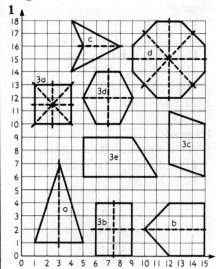

130

2a (1,1); (5,1); (3,7); (1,1)

  b (12,0); ((15,0); (15,4);
    (12,4); (10,2); (12,0)

  c (4,14); (8,16); (4,18);
    (5,16); (4,14)

  d (11,12); (13,12); (15,14);
    (15,16); (13,18); (11,18);
    (9,16); (9,14); (11,12)

3a Square
  b Rectangle
  c Parallelogram
  d Hexagon
  e Quadilateral

Page 57
1a b

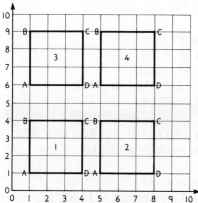

  c Square, **d** Area = 9 cm²

2a Activity **b** Square.
  c Area = 9 cm²
  d It has moved 4 cm to the right.

3a A(1,6); B(1,9); C(4,9);
    D(4,6); A(1,6)
  b Activities **c** Square,
  d Area = 9 cm²
  e It has moved 5 cm upwards

4a A(5,6); B(5,9); C(8,9);
    D(8,6); A(5,6)
  b Activities **c** Square,
  d Area = 9 cm²
  e It has moved 4 cm to the right
    and 5 cm upwards

Page 58
1a, b

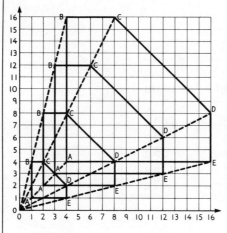

  c 7 cm²

2a Activity, **b** 63 cm², **c** 9,
d, e Activities
  f All the dashed lines end at (0,0)

3a A(2,2); B(2,8); C(4,8);
    D(8,4); E (8,2); A(2,2)
  b Activities
  c The new points all lie on the
    dashed lines continued to the
    right.
  d Area = 28 cm²,
  e Multiplying the number by 2
    multiplies the area by 4.

4a A(4,4); B(4,16); C(8,16);
    D(16,8); E(16,4) A(4,4)
  b Multiplying the numbers by 4,
    multiplies the area by 16.
  c The dashed lines meet with the
    points A,B,C,D and E of the
    new shape.

Page 59
1a, b, c Activities
  d (0,0); (2,2); (3,3); (5,5);
    (6,6); (7,7); (8,8); (10,10)

2

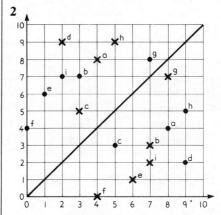

3a, b, c, d Activities

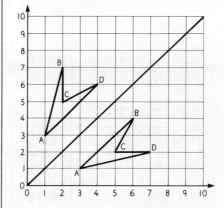

  e A(1,3), B(2,7), C(2,5),
    D(4,6), A(1,3)

4 The order of each pair of
  coordinates is reversed.

Chapter 13: Weight

Page 60

1 95 g

2a

	Gross weight	Nett weight	Weight of tin
peaches	1025 g	822 g	203 g
carrots	520 g	425 g	95 g
soup	490 g	435 g	55 g
beans	734 g	576 g	158 g
salmon	260 g	213 g	47 g

b 3029 g or 3·029 kg   **c** 203 g

3a 5 g

b

		Gross weight	Nett weight	Weight of tin
A	Chocolates	150 g	120 g	30 g
B	Jelly	200 g	180 g	20 g
C	Crunchies	460 g	375 g	85 g
D	Cornflakes	610 g	500 g	110 g
E	Jam	720 g	454 g	266 g
F	Pears	1025 g	792 g	233 g

c 20 g (B Jelly)

Page 61

1a 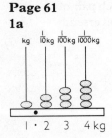   **b** 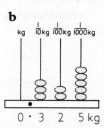   **e** 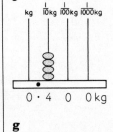   **f**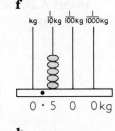

c 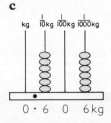   **d** 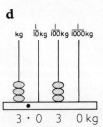   **g** 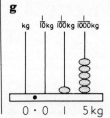   **h**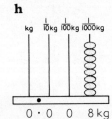

2a

	c
0·900 kg	0·300 kg
0·400 kg	0·080 kg
0·060 kg	0·040 kg
0·050 kg	0·027 kg
0·030 kg	0·002 kg

b	d
0·200 kg	0·750 kg
0·090 kg	0·725 kg
0·070 kg	0·720 kg
0·060 kg	0·700 kg
0·019 kg	0·500 kg

3a 1440 kg or 1·44 kg,
 b 439 kg or 0·439 kg,
 c 449 kg or 0·449 kg,
 d 3395 kg or 3·395 kg

Page 62

1a E 2·500 kg
 b H 2 kg 27 g
 c G, E, H, A, D, F, B, C
 d 22·473 kg

2a ≃ 2 kg   **b** ≃ 2 kg   **c** ≃ 4 kg
 d ≃ 1 kg   **e** ≃ 1 kg

3a ≃ 12 kg
 b Total weight = 11·827 kg,
 c Difference = 0·173 kg or 173 g

Page 63

1 4685 g

2 Difference 1·631 kg or 1631 g

3 2·310 kg

4 Tin of tomatoes, by 14 g.

Chapter 14: Fractions 1

Page 64

1a $\frac{2}{6}$ $\frac{3}{9}$ $\frac{5}{15}$ **b** Yes

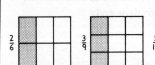

2a $7 \times 1 = 7$, **d** $\frac{1}{2} \times 1 = \frac{1}{2}$,
 b $10 \times 1 = 10$, **e** $\frac{1}{5} \times 1 = \frac{1}{5}$,
 c $1010 \times 1 = 1010$, **f** $\frac{1}{10} \times 1 = \frac{1}{10}$

3a $7 \div 1 = 7$, **d** $\frac{1}{2} \div 1 = \frac{1}{2}$,
 b $10 \div 1 = 10$, **e** $\frac{1}{5} \div 1 = \frac{1}{5}$,
 c $1010 \div 1 = 1010$, **f** $\frac{1}{10} \div 1 = \frac{1}{10}$

Page 65

1a $\frac{3}{4} = \frac{6}{8}$, **h** $\frac{12}{16} = \frac{3}{4}$, **o** $\frac{5}{8} = \frac{15}{24}$,
 b $\frac{8}{12} = \frac{2}{3}$, **i** $\frac{3}{10} = \frac{9}{30}$, **p** $\frac{32}{40} = \frac{4}{5}$,
 c $\frac{4}{5} = \frac{16}{20}$, **j** $\frac{22}{33} = \frac{2}{3}$, **q** $\frac{5}{8} = \frac{30}{48}$,
 d $\frac{21}{28} = \frac{3}{4}$, **k** $\frac{2}{5} = \frac{14}{35}$, **r** $\frac{35}{50} = \frac{7}{10}$,
 e $\frac{2}{3} = \frac{18}{27}$, **l** $\frac{12}{28} = \frac{3}{7}$, **s** $\frac{7}{8} = \frac{49}{56}$,
 f $\frac{70}{100} = \frac{7}{10}$, **m** $\frac{1}{2} = \frac{16}{32}$, **t** $\frac{27}{45} = \frac{3}{5}$,
 g $\frac{3}{5} = \frac{9}{15}$, **n** $\frac{36}{42} = \frac{6}{7}$,

Page 66

1a Fraction shaded $\frac{6}{15}$
  Fraction family $\frac{2}{5}$
 b Fraction shaded $\frac{2}{10}$
  Fraction family $\frac{1}{5}$
 c Fraction shaded $\frac{5}{10}$
  Fraction family $\frac{1}{2}$
 d Fraction shaded $\frac{5}{25}$
  Fraction family $\frac{1}{5}$
 e Fraction shaded $\frac{6}{12}$
  Fraction family $\frac{1}{2}$
 f Fraction shaded $\frac{3}{9}$
  Fraction family $\frac{1}{3}$

2a $\frac{4}{7} + \frac{2}{7} = \frac{6}{7}$
 b $\frac{3}{8} + \frac{4}{8} = \frac{7}{8}$
 c $\frac{1}{6} + \frac{5}{6} = 1$
 d $\frac{2}{9} + \frac{4}{9} + \frac{1}{9} = \frac{7}{9}$
 e $\frac{2}{11} + \frac{7}{11} = \frac{9}{11}$
 f $\frac{4}{13} + \frac{1}{13} + \frac{3}{13} = \frac{8}{13}$

Page 67

1a $\frac{5}{8} - \frac{3}{8} = \frac{2}{8} = \frac{1}{4}$
 b $\frac{5}{6} - \frac{1}{6} = \frac{4}{6} = \frac{2}{3}$
 c $\frac{7}{12} - \frac{5}{12} = \frac{2}{12} = \frac{1}{6}$
 d $\frac{4}{7} - \frac{2}{7} = \frac{2}{7}$
 e $\frac{7}{11} - \frac{2}{11} - \frac{1}{11} = \frac{4}{11}$
 f $\frac{7}{10} - \frac{3}{10} = \frac{4}{10} = \frac{2}{5}$
 g $\frac{7}{9} - \frac{2}{9} = \frac{5}{9}$
 h $\frac{11}{12} - \frac{10}{12} = \frac{1}{12}$
 i $\frac{6}{9} - \frac{2}{9} = \frac{4}{9}$

2a $\frac{7}{9} - \frac{2}{9} + \frac{4}{9} = \frac{9}{9} = 1$
 b $\frac{17}{20} - \frac{11}{20} = \frac{6}{20} = \frac{3}{10}$
 c $\frac{5}{12} + \frac{6}{12} - \frac{8}{12} = \frac{3}{12} = \frac{1}{4}$
 d $\frac{5}{8} - \frac{1}{8} = \frac{4}{8} = \frac{1}{2}$

3a $\frac{7}{12}$, **e** $\frac{3}{16}$, **h** $\frac{11}{16}$,
 b $\frac{13}{16}$, **f** $\frac{12}{15} = \frac{4}{5}$, **i** $\frac{5}{6}$,
 c $\frac{7}{12}$, **g** $\frac{5}{10} = \frac{1}{2}$, **j** $\frac{11}{16}$
 d $\frac{1}{10}$,

Page 68

1a $\frac{7}{10}$, **d** $\frac{1}{12}$, **g** $\frac{29}{30}$, **j** $\frac{5}{36}$,
 b $\frac{1}{12}$, **e** $\frac{17}{24}$, **h** $\frac{5}{24}$, **k** $\frac{13}{20}$,
 c $\frac{13}{15}$, **f** $\frac{3}{20}$, **i** $\frac{17}{30}$, **l** $\frac{3}{20}$

Chapter 15: Graphs

Page 69

1a 25, **b** Friday, **c** Tuesday,
 d 430, **e** $\frac{3}{4}$, **f** $\frac{2}{10}$ or $\frac{1}{5}$,
 g £26·80

2a

Attendances	Monday	Tuesday	Wednesday	Thursday	Friday
Adults	50	47	59	58	68
Children	30	28	31	27	32

b £114·20

3 Graph showing attendances for Class 4

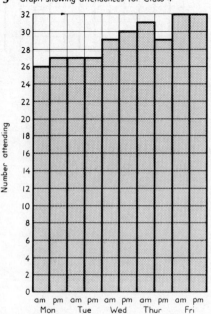

Page 70

1

Pet		Total
Dogs	ɪɪɪ ɪɪɪ ɪɪɪ ɪɪɪ ɪɪɪ ɪɪɪ ɪɪɪ ɪɪɪɪ	39
Cats	ɪɪɪ ɪɪɪ ɪɪɪ ɪɪɪ ɪɪɪ ɪɪɪ ɪɪɪ ɪɪɪ ɪɪɪ	43
Budgerigars	ɪɪɪ ɪɪɪ	8
Tortoises	ɪɪɪ	3
Rabbits	ɪɪɪ ɪɪɪ ɪɪ	12
Hamsters	ɪɪɪ ɪ	6
Parrots	ɪɪ	2
Grass snakes	ɪ	1

2 Pets owned by 3rd year pupils

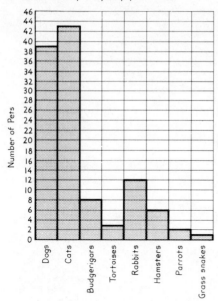

3, 4, 5 Open answers

Page 71

1 Highest temperature: July
Lowest temperature: January

2 16°C

3 December and February,
November and March.

Average maximum temperatures in Limnos

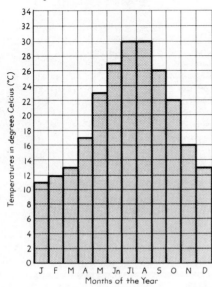

Average maximum temperatures in Naxos

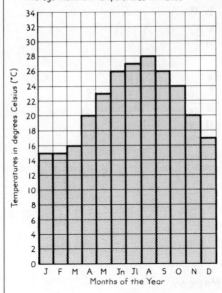

Average maximum temperatures in Zakynthos

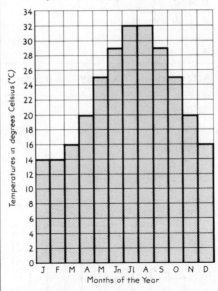

5 Limnos 19°C, Naxos 13°C,
Zakynthos 18°C

6 January, February, March,
April, October, November,
December

7 11°C warmer

Page 72

1 4 squares

2 25p

3 5 squares

4 2 mults

5a 24 mu,   **d** 28 mu,   **g** 34 mu,
b 40 mu,   **e** 12 mu,   **h** 22 mu
c 20 mu,   **f** 30 mu,

6a £2·50,   **d** £5·25,   **g** £4·62½,
b £3·75,   **e** £4·75,   **h** £3·12½
c £4·50,   **f** £5·75,

Page 73

1a £1 = 10 mu,   **d** £4 = 40 mu,
b £2 = 20 mu,   **e** £5 = 50 mu,
c £3 = 30 mu,   **f** £0 = 0 mu

2, 3

Graph for changing £'s to 'mults'

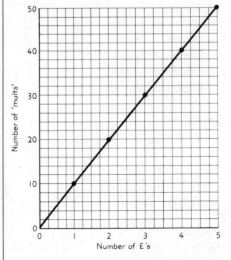

4	£	1	3	4	5	1·50	3·50	2·75	3·25	4·75	3·75
	mults	10	30	40	50	15	35	$27\frac{1}{2}$	$32\frac{1}{2}$	$47\frac{1}{2}$	$37\frac{1}{2}$

5	£	7	9	13	15	24	30
	mults	70	90	130	150	240	300

6 £1·50

7 30p

8 80p

9 £232, £185·60, £46·40

Page 74
1, 2, 3 Activities

4a $1 + 2 = 3$,   **f** $4\frac{1}{2} + 3\frac{1}{2} = 8$,
b $3 + 5 = 8$,   **g** $7 + 2\frac{1}{2} = 9\frac{1}{2}$,
c $7 + 2 = 9$,   **h** $3\frac{1}{2} + 5 = 8\frac{1}{2}$,
d $3\frac{1}{2} + 2\frac{1}{2} = 6$,   **i** $6\frac{1}{2} + 1 = 7\frac{1}{2}$,
e $6\frac{1}{2} + \frac{1}{2} = 7$,   **j** $5\frac{1}{2} + 2\frac{1}{2} = 8$

5 C

6a $7 - 5 = 2$,   **e** $8\frac{1}{2} - 3\frac{1}{2} = 5$,
b $8 - 2 = 6$,   **f** $6\frac{1}{2} - 5\frac{1}{2} = 1$,
c $6 - 4 = 2$,   **g** $8 - 3\frac{1}{2} = 4\frac{1}{2}$,
d $7\frac{1}{2} - 2\frac{1}{2} = 5$,   **h** $6\frac{1}{2} - 4 = 2\frac{1}{2}$

Chapter 16: Multiplication 2

Page 75
1a $4 \times 10 = 40$,   **e** $6 \times 10 = 60$,
b $3 \times 10 = 30$,   **f** $9 \times 10 = 90$,
c $8 \times 10 = 80$,   **g** $7 \times 10 = 70$,
d $5 \times 10 = 50$,   **h** $10 \times 10 = 100$

2a 120,   **e** 260,   **i** 2030,
b 150,   **f** 400,   **j** 1000,
c 210,   **g** 600,   **k** 1300,
d 190,   **h** 910,   **l** 7000,

3a 80,   **e** 840,   **i** 2610,
b 350,   **f** 960,   **j** 600,
c 720,   **g** 2240,   **k** 1600,
d 360,   **h** 2300,   **l** 5600

4a 400,   **d** 2700,   **g** 4000,
b 600,   **e** 2300,   **h** 7000,
c 1300,   **f** 8900,

5a 800,   **d** 25600,   **g** 20300,
b 3500,   **e** 13800,   **h** 19000
c 5200,   **f** 20500,

Page 76
1a 156,   **c** 180,   **e** 342
b 182,   **d** 240,

2a 168,   **e** 234,   **h** 272,
b 210,   **f** 256,   **i** 288,
c 195,   **g** 224,   **j** 361
d 196,

Page 77
1a 391,   **e** 609,   **i** 2666,
b 630,   **f** 1288,   **j** 2958,
c 703,   **g** 1924,   **k** 4278,
d 575,   **h** 1564,   **l** 4602

Page 78
1a 816,   **b** 1014,   **c** 3068,
d 4977

2a 1591   **e** 1802,   **i** 8178,
b 1551,   **f** 4307,   **j** 3150,
c 3752,   **g** 5440,   **k** 3162
d 841,   **h** 2535,

3a 448 children,   **b** 576 tickets,
c 1161,   **d** £7·70,
e 864 nails,   **f** 609 m²,
g 121, 144, 169, 196, 225, 256, 289, 324, 361, 400   **h** 1225, 1224, 1221, 1216, 1209, 1200, etc.

Chapter 17: Shape 2

Page 79
1, 2, 3 Activities

4a 70°,   **c** 100°,   **e** 65°,
b 60°,   **d** 65°,   **f** 77°

Page 80

	Square	Rhombus	Rectangle	Parallelogram
Four sides of a quadrilateral?	Yes	Yes	No	No
Four angles of a quadrilateral equal?	Yes	No	Yes	No
Diagonals same length?	Yes	No	Yes	No
Do diagonals cut each other in half?	Yes	Yes	Yes	Yes
Do diagonals cross at right angles?	Yes	Yes	No	No
Are diagonals axes of symmetry?	Yes	Yes	No	No
Are the angles of the quadrilateral cut in half by the diagonals?	Yes	Yes	No	No

Page 81

1

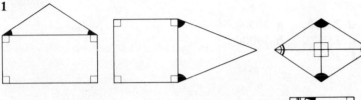

 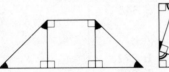

2a 120°, **c** 55°, **e** 110°, **g** 50°,
b 150°, **d** 35°, **f** 40°, **h** 80°

Page 82 Activities

Chapter 18: Division 2

Page 83
1a 400, **d** 2700, **g** 7300,
b 900, **e** 3900, **h** 8600
c 1200, **f** 4800,

2a 134
4)536
   − 400 │ 100(4)
    136
   − 120 │ 30(4)
     16
   − 16 │   4(4)
      0 │ 134(4)

b 148 r 5
6)893
   − 600 │ 100(6)
    293
   − 240 │ 40(6)
     53
   − 48 │   8(6)
      5 │ 148(6)

c 199 r 1
3)598
   − 300 │ 100(3)
    298
   − 270 │ 90(3)
     28
   − 27 │   9(3)
      1 │ 199(3)

Page 84
1a 188, **d** 156, **g** 121 r 7,
b 146 r 1, **e** 118 r 5, **h** 109
c 134 r 1, **f** 141,

2a 234
4)936
   − 800 │ 200(4)
    136
   − 120 │ 30(4)
     16
   − 16 │   4(4)
      0 │ 234(4)

b 473 r 1
2)947
   − 800 │ 400(2)
    147
   − 140 │ 70(2)
      7
   −   6 │   3(2)
      1 │ 473(2)

c 278
3)834
   − 600 │ 200(3)
    234
   − 210 │ 70(3)
     24
   − 24 │   8(3)
      0 │ 278(3)

d 249 r 2, **h** 356 r 1,
e 227 r 3, **i** 245 r 1,
f 413, **j** 291 r 1,
g 210, **k** 249 r 3

Page 85

1a   12 r 13

$$15\overline{)193}$$
$$-150 \mid 10(15)$$
$$\quad 43 \mid$$
$$-30 \mid 2(15)$$
$$\quad 13 \mid 12(15)$$

b   12 r 14

$$17\overline{)218}$$
$$-170 \mid 10(17)$$
$$\quad 48 \mid$$
$$-34 \mid 2(17)$$
$$\quad 14 \mid 12(17)$$

c   12 r 6

$$18\overline{)222}$$
$$-180 \mid 10(18)$$
$$\quad 42 \mid$$
$$-36 \mid 2(18)$$
$$\quad 6 \mid 12(18)$$

d 12 r 5,   **h** 13 r 3,
e 12 r 6,   **i** 11 r 12,
f 12 r 5,   **j** 12,
g 11 r 16,   **k** 17 r 3

2a 24 stamps, **d** 29 r 8, **g** 42 r 18,
b 31 r 12,   **e** 51,   **h** 52 r 1
c 26 r 9,   **f** 32 r 8,

3a 24 stamps, **b** 28 children,
c 24 shelves,
d 31 r 11
e 31 bottles

Page 86

1 Activity

2a 01·00, **g** 21·30, **l** 15·05,
b 13·00, **h** 10·15, **m** 23·59,
c 12·00, **i** 22·25, **n** 00·01,
d 05·00, **j** 20·45, **o** 12·05,
e 17·00, **k** 14·35, **p** 14·50,
f 24·00,

Page 87

1a 6 hours 20 minutes,
b 9 hours 25 minutes,
c 4 hours 20 minutes,
d 9 hours 15 minutes,
e 5 hours 8 minutes

2a 6 hours 30 minutes,
b 4 hours 45 minutes

3a 5 hours 48 minutes,
b 20.05

Page 88

1a 4 hours,
b 1 hour 30 minutes,
c 7 hours,
d 17 hours,
e 9 hours 30 minutes,
f 6 hours 15 minutes,
g 5 hours 13 minutes,
h 4 hours 16 minutes,
i 27 minutes

2a 6 hours 45 minutes,
b 6 hours 30 minutes,
c 9 hours 30 minutes,
d 35 minutes,
e 12 hours 35 minutes,
f 6 hours 25 minutes,
g 9 hours 52 minutes,
h 3 hours 17 minutes,
i 9 hours 38 minutes

3a No,
b 12 minutes, 4 minutes,
    2 minutes, 7 minutes
c 0758
d 1 hour 25 minutes,
e 16·42

4 Activity

Page 89

1a 7 hours 15 minutes,
b 17 hours,
c 64 hours 40 minutes,
d 28 hours 30 minutes,
e 31 hours,
f 44 hours 12 minutes,
g 61 hours 15 minutes,
h 19 hours 28 minutes,
i 79 hours 44 minutes,

2a 6 hours 40 minutes,
b 33 hours 20 minutes,

3 47 hours

4a 10 hours 15 minutes,
b 61 hours 30 minutes

Page 90

1 B $\frac{15}{8} = 1\frac{7}{8}$, C $\frac{13}{8} = 1\frac{5}{8}$,
    D $\frac{17}{8} = 2\frac{1}{8}$, E $\frac{9}{8} = 1\frac{1}{8}$

2a $1\frac{4}{5} = \frac{9}{5}$,   **g** $3\frac{4}{5} = \frac{19}{5}$,
b $2\frac{2}{3} = \frac{8}{3}$,   **h** $4\frac{1}{4} = \frac{17}{4}$,
c $2\frac{4}{5} = \frac{14}{5}$,   **i** $4\frac{2}{3} = \frac{14}{3}$,
d $3\frac{1}{4} = \frac{13}{4}$,   **j** $4\frac{7}{8} = \frac{39}{8}$,
e $3\frac{2}{3} = \frac{11}{3}$,   **k** $5\frac{7}{10} = \frac{57}{10}$,
f $2\frac{7}{8} = \frac{23}{8}$,   **l** $6\frac{2}{3} = \frac{20}{3}$

3a $3\frac{1}{2}$,   **e** $5\frac{2}{5}$,   **i** $5\frac{3}{8}$,
b $4\frac{2}{3}$,   **f** $4\frac{1}{6}$,   **j** $7\frac{1}{9}$,
c $2\frac{3}{5}$,   **g** $3\frac{3}{8}$,   **k** $6\frac{7}{8}$,
d $4\frac{3}{4}$,   **h** $4\frac{1}{10}$,   **l** $8\frac{4}{9}$

Page 91

1a $4\frac{7}{12}$,   **d** $4\frac{13}{20}$,   **g** $5\frac{19}{24}$,
b $5\frac{4}{5}$ or $5\frac{8}{10}$, **e** $7\frac{19}{24}$,   **h** $6\frac{11}{12}$
c $5\frac{11}{15}$,   **f** $5\frac{11}{20}$,

2a $6\frac{5}{12}$,   **e** $6\frac{5}{24}$,   **i** $6\frac{7}{24}$,
b $5\frac{1}{10}$,   **f** $9\frac{9}{20}$,   **j** $7\frac{19}{40}$,
c $4\frac{1}{24}$,   **g** $6\frac{7}{15}$,   **k** $7\frac{1}{36}$,
d $7\frac{3}{20}$,   **h** $8\frac{1}{3}$ or $8\frac{4}{12}$, **l** $10\frac{13}{40}$

Page 92

1a $1\frac{3}{4}$,   **e** $2\frac{13}{15}$,   **i** $3\frac{11}{15}$,
b $1\frac{5}{6}$,   **f** $1\frac{17}{24}$,   **j** $\frac{17}{18}$,
c $1\frac{19}{24}$,   **g** $1\frac{9}{10}$,   **k** $1\frac{29}{36}$,
d $1\frac{19}{20}$,   **h** $1\frac{2}{3}$ or $1\frac{4}{6}$, **l** $\frac{19}{24}$

Page 93

1a 31 mm, 3·1 cm
b 39 mm, 3·9 cm
c 72 mm, 7·2 cm
d 117 mm, 11·7 cm
e 136 mm, 13·6 cm

Page 94

1a 20 mm = 2 cm,
b 70 mm = 7 cm,
c 7 mm = 0·7 cm,
d 30 mm = 3 cm,
e 80 mm = 8 cm,
f 85 mm = 8·5 cm,
g 3 mm = 0·3 cm,
h 100 mm = 10 cm,
i 92 mm = 9·2 cm,
j 1 mm = 0·1 cm,
k 120 mm = 12 cm,
l 135 mm = 13·5 mm

2a

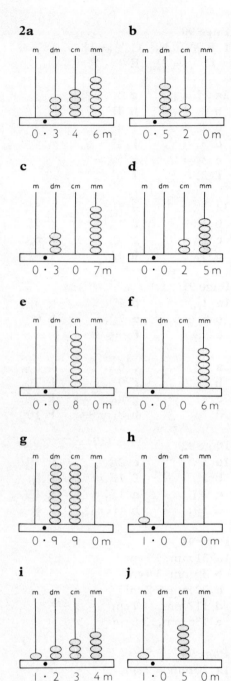

a 0 · 3 4 6 m

b 0 · 5 2 0 m

c 0 · 3 0 7 m

d 0 · 0 2 5 m

e 0 · 0 8 0 m

f 0 · 0 0 6 m

g 0 · 9 9 0 m

h 1 · 0 0 0 m

i 1 · 2 3 4 m

j 1 · 0 5 0 m

Page 95

1a Activity,
  b Perimeter ≃ 240 mm or
  24 cm   **c** 232 mm

2 A Perimeter ≃ 160 mm
    Perimeter = 158 mm, 0·158 m
  B Perimeter ≃ 200 mm
    Perimeter = 196 mm, 0·196 m
  C Perimeter ≃ 240 mm
    Perimeter = 226 mm, 0·226 m
  D Perimeter ≃ 200 mm
    Perimeter = 185 mm, 0·185 m

3a A Right angled triangle
   B Delta
   C Parallelogram
   D Pentagon

3b 70 mm, 0·070 m
  c As all sides are equal multiply
   the length of a side by 5.

Page 96
1 125 mm

2 Perimeter ≃ 1020 mm or 1·020 m
  Perimeter = 1032 mm or 1·032 m

3 Activity

4 Open answer

5 Activity

Index

Nuffield Maths 5–11

Nuffield Maths 1   Teachers' Handbook
Contents

1 **Sets and relations (N1)**
N1:1 Relations (different types of correspondence)
N1:2 Early sorting experiences—leading to sets
N1:3 Sorting into subsets

2 **Matching (N2)**
N2:1 Matching to find equivalent sets
N2:2a Two non-equivalent sets
N2:2b Three non-equivalent sets
N2: Pictorial representation and semi-tallying

3 **Counting and numerals (N3)**
N3:1 Counting
N3:2 Matching a number symbol to a set
N3:3 Introducing number words
N3:4 The empty set
N3:5 Conservation of number

4 **Ordering (N4)**
N4:1 Ordinal numbers
N4:2 Putting non-equivalent sets in order
N4:3 Signs 'is greater than', 'is less than'
N4:4 Tallying and pictorial representation

5 **Towards addition (N5)**
N5:1 Composition of small numbers
  1a bead bag 1b duck pond 1c rods or strips and frames
N5:2a Addition of two disjoint sets; putting sets together
N5:2b Addition of two disjoint sets, using structured apparatus
N5:3 Recording addition by mapping

6 **Addition to 10 (N6)**
N6:1 Number bonds up to 10
N6:2 Counting on
N6:3 Patterns in simple addition
N6:4 Picture problems—additions
N6:5 Introduction of addition sign and vertical addition
N6:6 The addition square

7 **Length (L1)**
L1:1 Descriptive language
L1:2 Comparing two unequal lengths
L1:3a Matching lengths: matching two objects of about the same length
L1:3b Matching lengths: using several objects to 'make up' a length
L1:4 Ordering
L1:5 Measuring with repeated units
L1:6 Using limb measures

8 **Shape and space (S1)**
S1:1 Awareness of shape and space, extension of vocabulary
S1:2 Early sorting of 3-D shapes
S1:3 Early sorting of 2-D shapes
Appendix

9 **Weighing (W1)**
W1:1 Descriptive language—heavy and light
W1:2 Comparing—heavier than and lighter than
W1:3 Balancing
W1:4 Ordering

10 **Time (T1)**
T1:1 Association—matching events to daytime or night-time
T1:2 Putting time sequence in order
T1:3a Comparisons: fast and slow
T1:3b Comparisons: timing
T1:4 Graphs and charts

11 **Money (M1)**
M1:1 Recognition of 1p, 2p, 5p, 10p coins
M1:2 Comparison of amounts of money (a) by matching (b) by totals
M1:3 Using coins to make amounts up to 10p
M1:4 Early stages of shopping
Appendix: Suggestions for shops in schools

12 **Capacity (C1)**
C1:1 Full, empty and half empty
C1:2 Which holds more? (visual judgement)
C1:3 Finding the capacity by counting
C1:4 Sorting containers
C1:5a Comparing by emptying
C1:5b Comparing by filling

Nuffield Maths 2   Teachers' Handbook
Contents

1 **Early stages of subtraction (N7)**
N7:1 Finding the difference
N7:2 Counting back
N7:3 Taking away
N7:4 Recording subtraction
N7:5 Practice sheets

2 **A first look at place value (N8)**
N8:1 Early grouping activities and games
N8:2 Grouping using cubes, etc.
N8:3 Grouping in tens

3 **Addition to 20 (N9)**
N9:1 Number bonds up to 20
N9:2 Counting on
N9:3 Ways of recording

4 **Subtraction involving numbers up to 20 (N10)**
N10:1 Difference by matching and counting
N10:2 Subtraction by counting back
N10:3 Taking away
N10:4 Ways of recording

5 **Introducing multiplication (N11)**
N11:1 Recognising and counting equivalent sets
N11:2 Multiplication as repeated addition
N11:3 Arrays and the commutative law
N11:4 Activities and games for 'table facts' up to 30

6 **Introducing division (N12)**
N12:1 The sharing aspect of division
N12:2 The repeated subtraction aspect of division
N12:3 Division as the inverse of multiplication
N12:4 Remainders

7 **Length (L2)**
L2:1 Appreciating the need for a standard measure
L2:2 Introduction of the metre
L2:3 Comparison with a 10 cm rod (decimetre)
L2:4 Measuring in centimetres—straight and curved lines
L2:5 Personal measurements in m and cm

8 **Shape and Space (S2)**
S2:1 Sorting for shape and size
S2:2 Fitting shapes together
S2:3 Surfaces and faces
S2:4 Covering surfaces—leading to area
S2:5 First ideas of symmetry

9 **Weighing (W2)**
W2:1 Introduction of kilogram and ½ kilogram
W2:2 Using the kilogram and ½ kilogram
W2:3 Introduction of the 100 gram weight

10 **Time (T2)**
T2:1 Ways of measuring time
T2:2 Reading a dial
T2:3 Telling the time (hours, halves, quarters)
T2:4 Telling the time (five-minute intervals)
T2:5 Simple calculations involving time
T2:6 Other units of time

11 **Money (M2)**
M2:1 Reinforcement of coins up to 10p and introduction of 50p
M2:2 Breakdown of coins—equivalent values
M2:3 Making amounts up to 20p
M2:4 Addition—simple shopping bills
M2:5 Giving change and finding difference by counting on
M2:6 Subtraction by taking away

12 **Capacity (C2)**
C2:1 Introduction of the litre
C2:2 Comparing a litre with non-standard measure
C2:3 Introduction of ½ litre and ¼ litre
C2:4 Cubes, boxes and walls

Nuffield Maths 3 Teachers' Handbook
Contents

Nuffield Maths 4 Teachers' Handbook
Contents